GOD'S RESET

HOW A FAMILY SURVIVES THE GRAND SOLAR MINIMUM

By Hugh Simpson

Copyright December 2023

All Rights Reserved.

INTRODUCTION

As the front cover says the family in the story is fictional; however what I am calling the beginning of **GOD'S RESET** is a REAL event that occurs approximately every 350 years – **THE GRAND SOLAR MINIMUM!**

I was one of the first to write about the effects of Y2K after discovering two paragraph article in 1998. It led to my very successful **A Family Survival Manual for Y2K & Beyond** and now a 23 year career as a Preparedness and Survival Skills consultant having written eight books in the niche with the latest being **XtremePreparedness™ on Land & Sea** under my pen name MR Valentine exclusively for Amazon.

I have been a guest on the internationally syndicated Coast to Coast AM numerous times with the first in 1999 with the legendary late Art Bell. Recently I also guested on Night Dreams Talk Radio with Art's long time friend Gary Anderson. I was a guest columnist for **Survivors Edge**. My preparedness tactics and strategy for surviving the effects of Y2K led to appearances on Fox News and CNN.

I had hoped my current efforts to awake people to the potentially devastating impact of this current Grand Solar Minimum would work.

It has not and now we preparedness consultants and not those "wannabees" see that less than 2% are prepared to survive this potential **35 YEARS** event!

When you read about what has been going on with the US government's plan to protect the President and other important people from the effects, you can take it to the bank that it is REAL as one of my suppliers told me why they wanted his revolutionary power generating system OFF the general public market for THREE YEARS and paid him handsomely for the contract!

If you want to SURVIVE I beg you to read this novel as not just a hopefully inspiring story. I know it may seem long and drawn out; but if you follow every step of what this family does to SURVIVE you will have a better chance of SURVIVING! It is based on my decades of EXPERIENCE of helping others like YOU!

This is just Phase 1 OF **GOD'S RESET**! At the end of the book I will discuss what could be the FINAL RESET!

Hugh Simpson

https://www.xtremepreparedness.com

PROLOGUE
2026

Bob, his wife Dorothy and children Andy and Cindy have their hand cranked radio tuned into a news cast.

"Folks, for any of you able to hear me this will most likely be our last broadcast because our generators are down to the last bit of gas.

"If you are listening then you know the predicted Grand Solar Minimum by renowned solar physicist Dr. Zharkova is occurring.

"The entire power grid of the United States is down. The chances of restoring it are nil. If Dr. Zharkova has said along with her colleague the Preparedness and Survival Skills consultant this Grand Solar Minimum will last potentially 35 years, there will be no restoration of the grid in most of our live times.

"Of course with this occurrence there has been the "domino effect".

"The worst situation is that all smartphones are eventually going to run out of power along with tablets, laptops and desk tops unless you have a hand cranked generator, which of course you can most likely not purchase now. Those wearables that you relied on are connected to smartphones so AI or Artificial Intelligence is worthless.

"No power means no gas for vehicles and even the electric vehicles are not going to be of any use.

"No power has led to the closing down of ALL businesses including banks with their ATMs and cryptocurrency sites. No grocery or convenience stores and certainly no fast food restaurants.

Those who were relying on paying with cryptocurrencies are no longer able to do so and crypto mining is a process powered by the grid.

"Those who began to BARTER for ammo, whiskey, cigarettes, cigars, toilet paper, etc. are the new money kings, who took the money and bought RVs, sailboats, tiny homes on wheels and land in safe areas long before the RESET are most likely laughing at our STUPIDITY!

"The powerful generators used by hospitals are beginning to shut down and all ambulances are at a standstill along with other emergency vehicles.

In fact, the majority of police, fire and EMTs are either not able to get to their stations or have fled the area to safer ones like south Georgia, Florida, south Texas or the far western states like California and Arizona. Of course the WATER is running out in those two states even quicker from the worst drought in over a thousand years. No water from the dammed lakes will mean NO power too.

"With no protection of the especially large urban areas, the looting is epidemic. Grocery and convenience stores have been the first ones hit with smashed in glass doors and windows. These areas look like ghost towns.

"Folks, it seems to be true that the wealthy were preparing for this catastrophic event with their mega yachts and massive relocations to safe areas. It is ironic that the USA built a wall across the western states to keep illegals crossing over and now Americans are fleeing to Mexico, who has power. However Mexican military are lined up along the border turning these Americans away!

"The US Virgin Islands, Puerto Rico and other Caribbean countries have experienced problems with the new armada of yachts arriving on these islands. Same for Mexico and the majority of Central America.

"Chaos reigns wherever the effects of this Grand Solar Minimum is occurring. GOD help us all!"

"Praise the Lord we saw what was happening and relocated," says a relieved Bob.

His family nod their heads sadly in agreement.

Table of Contents

CHAPTER 1 June 21, 2025 Topeka, Kansas 10

CHAPTER 2 June 22, 2025 Topeka, Kansas 13

CHAPTER 3 June 22, 2025 Pentagon Bunker 15

CHAPTER 4 June 23, 2025 Topeka Kansas 21

CHAPTER 5 June 23, 2025 Topeka Kansas 24

CHAPTER 6 June 23, 2025 Topeka Kansas 26

CHAPTER 7 June 24, 2025 Topeka Kansas 30

CHAPTER 8 June 24, 2025 Topeka Kansas 32

CHAPTER 9 June 25, 2025 Topeka Kansas 34

CHAPTER 10 June 26, 2025 Pentagon Bunker 36

CHAPTER 11 June 26, 2025 Topeka, Kansas 41

CHAPTER 12 June 27, 2025 Pentagon Bunker 48

CHAPTER 13 June 27, 2025 Topeka Kansas 51

CHAPTER 14 June 28, 2025 Topeka Kansas 58

CHAPTER 15 June 28, 2025 Topeka Kansas 60

CHAPTER 16 June 30, 2025 Topeka Kansas 68

CHAPTER 17 July 1, 2025 Topeka Kansas 70

CHAPTER 18 July 2, 2025 Franklin, NC 76

CHAPTER 19 July 4, 2025 Jesup, Georgia 78

CHAPTER 20 July 5, 2025 Jesup, Georgia 80

CHAPTER 21 July 5, 2025 Pentagon Bunker 83

CHAPTER 22	July 8, 2025 Topeka, Kansas	85
CHAPTER 23	July 9, 2025 Topeka, Kansas	86
CHAPTER 24	July 10, 2025 Jacksonville, Florida	88
CHAPTER 25	July 11, 2025 Topeka, Kansas	91
CHAPTER 26	July 16, 2025 Topeka, Kansas	93
CHAPTER 27	July 17, 2025 Topeka, Kansas	95
CHAPTER 28	July 18, 2025 Pentagon Bunker	97
CHAPTER 29	August 1, 2025 Topeka, Kansas	102
CHAPTER 30	August 14, 2025 Topeka, Kansas	104
CHAPTER 31	August 15, 2025 Topeka, Kansas	105
CHAPTER 32	August 16, 2025 Jesup, Georgia	107
CHAPTER 33	August 17, 2025 Pentagon Bunker	108
CHAPTER 34	August 17, 2025 Jesup, Georgia	110
CHAPTER 35	August 17, 2025 The White House	112
CHAPTER 36	August 17, 2025 Jesup, Georgia	114
CHAPTER 37	August 31, 2025 Jesup, Georgia	116

CHAPTER 1

June 21, 2025 Topeka, Kansas

"Sure is staying awfully cold for this time of the year, Dad," says Andy as he prepares to take the cows into the barn.

"Yes, I agree and now I'm beginning to believe that guy I heard on Coast to Coast AM might be on to something," says Andy's dad, Bob.

"What did you hear?" inquires Andy as he shuts the barn door.

"Let's head to the house and a hot cup of coffee," suggest Bob.

"Sounds good to me," says Andy as they head down the path into the old farm house. "I can't believe it is the first day of summer and barely above freezing."

Yes, it is actually June 21, 2024 and the temperature in Topeka Kansas is 54 degrees at 6PM. The normal temperature for the first day of summer even at 6PM is 70 degrees not 54 degrees.

People are paying more attention to these abnormal lows they are experiencing especially as farmers. This is the area of the United States that has traditionally been called "the bread basket" and these low temperatures are not boding well for a crop growing season of corn, wheat and even vegetables.

"So what did you hear on your Coast to Coast AM besides new sightings of Bigfoot or the Loch Nest Monster," teases Andy, who knows his Dad is a regular listener to the host George.

"This time it was this Preparedness and Survival Skills consultant, who had been previously a Post Newsweek TV Group investigative reporter like George," begins Bob. I almost turned him off but decided that this former investigative reporter might be a different change of pace than the usual guests.

"Actually he began to learn about preparedness and survival after he learned of an impending event called Y2K that could have had catastrophic effects. You were to young to know about this."

"Yeah, I have learned about Y2K in my computer science course and how for some stupid reason the computer geeks back then had not prepared for code to go past 2000," shares Andy. "What were they thinking?"

"Yes, and as this guest explained this event actually went right down to the bewitching hour with computer experts desperately fixing code before the beginning of the new Millennium," explains Bob.

"He went on to write and publish his first book **A Family Survival Manual for Y2K & Beyond**, continues Bob. "It led to appearances on Fox News, CNN and becoming a guest columnist for **Survivors Edge**. What he did not share with his initial interview with the original Coast to Coast AM founder the late Art Bell was what he read during his research about Y2K. That Art Bell was a legend and even had a guest appearance on the **X-Files** that I watched every Friday night. It was the **X-Files** that led to my interest in aliens, Bigfoot and the Loch Nest Monster."

"So what did the guest share with George when you heard him?" questions Andy.

"Well he found this article by two former Soviet Union astrophycists discussing an upcoming event that could freeze a substantial amount of the planet for literally decades," explains Bob as he prepares for coffee refills. "Then the guest is watching YouTube and sees this workshop being conducted by a former Soviet Union astrophysicist Dr. Zharkova, who had left the Ukraine to teach in the United Kingdom. She is also the foremost solar physicist and has developed an algorithm that she used to go back 10,000 years looking to the reoccurring of what are called Grand Solar Minimums. She found these occur approximately every 350 years and the last one in Europe wiped out 25% of the population by both starvation and freezing to death! In addition it created a Mini Ice Age."

"WOW! Are you serious?" says a wide eyed Andy. A Mini Ice Age. How long did it last?"

"According to Dr. Zharkova these Grand Solar Minimums last close to 35 years," explains Bob. "She even had another solar physicist confirm her algorithm."

"Do you think that this abnormal freezing weather is the beginning of another Grand Solar Minimum?" questions a concerned Andy. "If it is, how are we going to survive?"

"Let's discuss this later as I hear your Mom and Cindy arriving back home," answers Bob. "Your sister does not need to hear this as she is only nine and it would cause her to have nightmares."

"Heck, I don't know how well I will sleep tonight," shares Andy.

CHAPTER 2

June 22, 2025 Topeka, Kansas

Next morning Bob goes to his weekly sit down with other local farmers to discuss what is on their minds. He has decided to keep his information on this Grand Solar Minimum to himself just like he has done with his interest in Bigfoot.

"Man was it cold last night," states Ed one of the local farmers. "I have never seen it this cold in the summer. Any of you guys have an explanation or hear something I might have missed?"

"Yeah you are right," says Frank, who is the one known to their group as the 'know it all'. "I heard that local weatherman say it was some kind of weather weirdness brought on by some system BUT it would not be anything to concern us."

"We knew you would have the answer, Frank," teases George.

"Yeah, it actually was kind of concerning to me before I heard that weather guy," admits Frank. "However now with him being an authority on the weather I am no longer concerned. I'm sure you guys would have been wondering if this could shorten our growing season."

They all agree and head out of the door of the local restaurant to begin another day of livelihood on their farms.

Later in the afternoon Andy notes that the temperature is even colder than the 54 degrees the same 6 PM yesterday.

"Dad, does it feel like it is even a little colder than yesterday at this time?" ask Andy.

They both check the temperature gage at the barn and see it is 52 degrees.

"You are right, Andy," says Bob. "I was at the meeting with the guys this morning and Frank said he heard the weatherman say it was some kind of unusual system but it would dissipate over the next few days."

"So it appears that it has nothing to do with this Grand Solar Minimum?" inquires Andy.

"Frank did not hear that weatherman mention it or he would have been telling all of us about it," laughs Bob.

"Yeah, you are always saying he is the 'know it all'," laughs Andy as they head back to the house.

Little do Andy, Bob or Frank REALLY know what is going on.

CHAPTER 3

June 22, 2025 Pentagon Bunker

Deep down in the Pentagon is a newsroom that was secretly created by members of the US Army Corps of Engineers after the 911 Pentagon attack.

Sitting at a magnificent teak table is the first female five star Chairwoman of Joint Chief of Staff surrounded by generals and an admiral from the Army, Navy, Air Force, Marines, SPACE Force and National Guard. Additionally several physicists and meteorologists are present.

The Chairwoman Admiral Hutchinson opens the meeting by saying:

"Ladies and gentlemen this meeting is TOP SECRET and no information to ANYONE excluding the Secretary of Defense and State; right-to-know members of Congress or the Senate; the Vice President and President is to be shared especially to the media. Is that understood?"

Nods of agreement are noted around the table.

She continues:

"I would like to open with our distinguished physicist, who happen to be in agreement with Dr. Zharkova of the United Kingdom on the impact of this current Grand Solar Minimum. Dr. Matheson will you open with your presentation?"

"Thank you Admiral, states Dr. Mathison. "Yes, I have been a long time admirer of Dr. Zharkova even before she immigrated from the former Soviet Union. She is a very consciousness and thorough scientist, who developed an algorithm that studied the occurrence of Grand Solar Minimums over a period of 10,000 YEARS. Not trusting just her algorithm she had another solar physicist confirm it. Dr. Zharkova found that these Grand Solar Minimums occur approximately every 350 years and last on the average of 35 years. The last one was in the late 1660s to 1760 time frame creating a Mini Ice Age as seen in this rendering."

"It was so severe that it caused **25%** of the European population to die of Arctic freezing temperatures and starvation."

General Davidson of the Marine Corp interrupts by saying:

"Are you truly saying **25%** of the European population died? If we were to use this same percentage for the current US population of 335,000,000 plus, that would mean we could see over **83,000,000** deaths? This number is staggering and mind numbing!"

"Yes, it is General," agrees Dr. Matheson. "The only good news about this current Grand Solar Minimum is that Dr. Zharkova does not see a Mini Ice Age this time; however we will soon see the Arctic freezing temperatures."

Admiral Stetson of the Navy speaks up:

"So what are the options for saving as many US citizens as possible?"

"I am going to defer now to my distinguished colleague Dr. Edwards," says Dr. Matheson.

"Thank you Dr. Matheson, begins Dr. Edwards. "Ladies and gentlemen my area of physics is geophysics and I have studied the geography of especially the UNDERGROUND area of the United States. As you might be aware, any underground area at least **10** feet below ground level is around **50** degrees even on a day with below freezing weather above."

General Davidson speaks up again:

"So this is at least some good news; however how many square miles of the USA is underground?

"Excellent question General," states Dr. Edwards. "Currently there are over **205,000** square miles of cave systems; however a certain percentage would not be currently accessible or habitable. There is other good news that I read in an article in **Survivors Edge** magazine written by a well-known preparedness consultant and author with a former background as an investigative reporter. He discussed a number of new places that would be the equivalent of being underground: subway stations, basements 10 feet below ground level especially of large complexes, sub level shopping centers like the famous one in Montreal, and airport facilities like the one in Denver that some believe is up to eight stories below ground level. Some conspiracy advocates claim there are thousands of miles of underground tunnels with traffic already moving through them. Finally you have Elon Musk and his company Bore that has attempted to move Los Angeles traffic underground."

National Guard General Jackson speaks up:

"Very interesting Dr. Edwards. As you are aware I am responsible for any potentially catastrophic events like hurricanes, earthquakes, and rioting that could effect our population. This Grand Solar Minimum seems to be one of those events. However it will require ordinary citizens becoming more than even serving our country's interest in both Iraq and Afghanistan. It will certainly appear to be a longer period of time. Therefore General Smith I am going to need your cooperation using US Army troops from Fort Campbell."

General Smith of the Army responds:

"You are so right General as we assisted your National Guard during the Atlanta Olympics especially after the bombing; however are you aware that these especially trained warriors could currently only handle up to **EIGHT** cities like Atlanta? I do note from this rendering that this Grand Solar Minimum may NOT effect the US Western states. Is that correct Dr. Edwards or Matheson?"

"That appears correct currently, general; however I'm sure you are aware of the current drought out there, which is considered to be the worst in **1500 YEARS** and we are already seeing more water level drop on the lakes out there greatly effecting the power grid with more blackouts," replies Dr. Matheson. "We have more people capable of leaving that area doing so; however the majority are most likely not aware of the current Grand Solar Minimum."

"A truly FUBAR nightmare scenario," grumbles General Davidson. "I would like to know how we have not been made aware of this sooner?"

"I attempted to make several members of Congress and the Senate aware of the current Grand Solar Minimum and their aides just offered their traditional lip service saying they appreciated my concern and would look into it," replies Dr. Matheson.

"Madame Chairwoman, I have heard reports that we have been sending personnel into US abandoned mines under the auspices of both DARPA and Homeland Security to renovate them and evidently make them inhabitable," states Admiral Stetson.

"That sure is news to me, Admiral," says the Chairwoman Admiral Hutchison. "Do you have further information?

"Yes, I do as I have actually heard this same preparedness and survival skills consultant during his interview as one of my aides listens regularly," shares Admiral Stetson. "He thought it would be something I might like to hear. I agreed and he arranged for me to hear the rebroadcast.

"I learned that one of the guest's suppliers had been approached by a subcontractor to I believe DARPA to talk to the inventor of a revolutionary 1600 watt hand cranked generator about using it in one of their projects. They wanted him to remove the sale of this generator off the public market for three years, which he agreed to do.

"He was curious as to what they were being used for. Evidently DARPA actually sent a representative to his manufacturing facility every week and one young representative told him about the abandoned mine project."

"So who and when does DARPA and or Homeland Security plan to share this information with?" asks General Jackson. "I would have expected them to share this with me."

"I certainly agree," says the Chairwoman. "I plan to find out about this secret project and what they plan to do with those abandoned mines. Right now I would like to hear from our meteorologists as to the current status of this Grand Solar Minimum."

"Thank you, Madame Chairwoman," replies one of the meteorologist. "I am Dr. Anderson of NOAA and it is Dr. Matheson, who made me aware of Dr. Zharkova's research. I must admit at first I thought it was just one of our numerous solar minimums that last usually a few years and not this current one. I began by listening to Dr. Zharkova's interview 2020 that even now has over **152,000** views . Please start the YouTube video."

Lights go down and the YouTube video starts:

https://youtu.be/M_yqIj38UmY

"As you will note Dr. Zharkova is a very thorough scientist and I wanted you to hear that she uses totally scientific research and NOT speculation," continues Dr. Anderson.

"So Dr. Anderson I heard in that interview that agencies like NOAA have been aware of this Grand Solar Minimum since before the 2020 interview," states General Davidson. "So why are we in this room just learning about this current Grand Solar Minimum and its potential devastating effects on vegetation as Dr. Zharkova points out?"

"As Dr. Matheson has pointed out I too reached out to Congressional and Senatorial aides I knew from previous opportunities to act as an expert and they all said thank you and would look into it," says Dr. Anderson.

"Yeah too damn concerned about upcoming elections," replies Admiral Stetson. "So now how do we handle this scenario without causing wide spread panic?"

"Good question Admiral," says the Chairwoman. "Personally I want to study this more and convene another meeting in the next week or so. Meanwhile I definitely do not want our media to learn of this meeting as it would certainly lead to the wide spread panic you mentioned.

"Also Dr. Anderson I would like to see if we can keep our meteorologists on local and national media from mentioning anything related to this Grand Solar Minimum. Just keep them reporting that it is short lived weather event not to worry about. So unless there is any further questions or discussion I declare this meeting closed."

All agree, gathering up papers and notes and head out of the bunker.

CHAPTER 4

June 23, 2025 Topeka Kansas

Like clockwork Andy and his Dad are herding the cattle into the barn.

Andy checks the temperature to seeing it is even lower than the night before.

"Dad, look at the temperature now," says Andy pointing to the thermometer. "It is lower than yesterday at 6PM."

Bob looks to see it is currently 51.

"Frank said that the weatherman said it could go lower over the short period of this system," reminds Bob. "However I am still concerned about what we are seeing."

"By the way I found that guy's book on Amazon and purchased it in Kindle format," explains Andy. "He was the one on Coast to Coast AM that you heard. It is entitled **XtremePreparedness™ on Land & Sea**. I have already started reading it and found the link to Dr. Zharkova's interview and after searching found a later one. She is definitely a thorough scientist and a genius in computer science."

"I would like to watch those interviews," replies Bob. "I want to hear her thoughts especially as related to farming."

"Actually she says in the more recent interview that she has watched with each successive year the shortening of the growing season. I also found her own website at https://www.solargsm.com and there is even more up to date information."

"So what does the author say we should do to survive this Grand Solar Minimum?" questions Bob.

"First we need to create a BUG-OUT BACKPACK for EVERY member of the family," explains Andy "You, Mom and myself would use regular ones and Cindy would use a slack pack. Yours and mine would be able to easily hold up to 35 pounds of items. Mom's would be in the range of 20 and probably 10 for Cindy."

"I definitely want to get these backpacks right away," says Bob.

"The author says not to get those flimsy Made in China ones as they don't have the strong straps that are needed for long hauls when a break could be catastrophic," shares Andy. "He recommends a Made in the USA company at https://www.battlelakeoutdoors.com who have been designing and manufacturing these STRONG STURDY ones for first responders for years.

Andy opens his smartphone, where he has stored the Kindle book to the page with the photo of the backpack and its contents.

"WOW, that is amazing what is contained in that backpack!" exclaims Bob. "I would never have guessed you could get all those items in one! Now I understand why you would NEVER use some flimsy one."

"Yes, it is definitely packed with everything one would need," agrees Andy. "By the way, the author shares in the book a description of each item and where to get them."

(At the end of the book I have an offer for you the purchaser of this book, where in two videos I will show you what currently is in both my slack pack and backpack and you will have an opportunity with only 10 participants to ask me questions on the slack pack and backpack.)

"Okay, Andy, you are in charge of researching what we see here and where to get them," states Bob. "Also find, where we can get these items as cheaply and quickly as possible."

"Got it moving ahead already as I have listed each item and the link to them here on my phone," says Andy. "So are we going to share this project with Mom and Cindy?"

"Let me handle that aspect," responds Bob.

CHAPTER 5

June 23, 2025 Topeka Kansas

Tom, the local TV meteorologist is sitting at his desk studying the information he has received from NOAA about this unusual weather system bringing such low temperatures on the third day of SUMMER.

"Susan, I just don't understand why we are experiencing these cold temperatures at this time," says Tom to his other meteorologist. "I know NOAA is sending us this information that this system will not last much longer, but my 'gut' feeling is that this system is way more than what we are being told."

"Tom, I tend to agree with you and I decided to do some investigating on this system," explains Susan. "I have in addition to a long time interest in meteorology had an interest in solar activity. So I started Googling and found something that occurred the last time 350 years ago called the Grand Solar Minimum as explained in a video by Dr. Zharkova, who is a well respected solar physicist and an authority on what she considers the new Grand Solar Minimum."

"Fascinating, go on," exclaims Tom.

"In her video she explains what causes the Grand Solar Minimum and what occurs during it. There is a substantial decrease in sunspot activity which leads to a COOLER temperature on the planet. The last Grand Solar Minimum actually created a Mini Ice Age; however Dr. Zharkova does not see this happening with the current one. Still, the temperatures will be very cold and will cause a potentially catastrophic shortened growing season. That really caught my attention since we are a primary growing area for corn, soybeans, wheat, etc. for the USA."

"I want to watch this video," says Tom. "I would like for you to do more research on this Grand Solar Minimum that the physicist speaks about. If this is the new Grand Solar Minimum, then we need to let our viewers know about it."

Susan begins to Google and finds more information on the current Grand Solar Minimum especially that Dr. Zharkova states that this one actually began around the late 2019 or early 2020. She learns that now will be the phase, when they will start experiencing the colder temperatures and that they could last for 35 years. Also she finds farmers in Canada already talking about what they are experiencing since 2019. They are saying there has been a distinct drop in temperatures not expected and very concerning.

Susan shares her research with Tom, who has watched the video.

"I have decided to lead my weather segment with a potential explanation of why we are seeing these unexpected low temperatures," states Tom. "So I want us to have the graphs we have found and especially this rendering to explain more about this Grand Solar Minimum. I have weighed the pros and cons of doing this and I think it is responsibility of us as meteorologists to present what we know."

"I support your decision 100%," affirms Susan.

CHAPTER 6

June 23, 2025 Topeka Kansas

While watching the six o'clock news out of Topeka, Andy, Bob, and Mom Dorothy see the meteorologist, Tom, "tease" his upcoming weather segment with reference to the unusually cold weather system.

"I wonder what he is going to say?" questions Andy.

"It sure will be interesting," replies Bob.

"Yes, the ladies in my sewing club have been discussing it too," says Dorothy.

Now the meteorologist Tom reappears and says in a few minutes he will be sharing the research he and his assistant meteorologist Susan have been doing all afternoon into what could be causing these continuing cold summer days and nights.

Andy looks over at his Dad Bob, who shrugs his broad shoulders.

Tom finishes his normal weather predictions and leads into the promised potential explanation with this:

"What you are looking at is a recent rendering Susan and I discovered of an event that occurred approximately 350 years ago creating a Mini Ice Age. It is called a Grand Solar Minimum. This is when our sun goes into a phase with a significant drop in sunspots as seen here directly from NASA," says Tom.

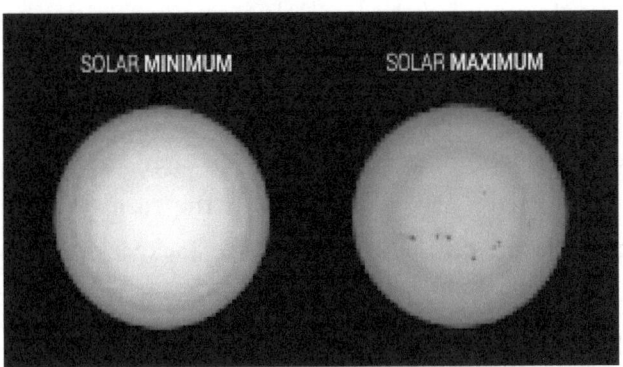

"As you will note there is a definite DECREASE in sunspots," continues Tom. "We know also that with the drop in sunspots there is a time of colder temperatures. Usually Solar Minimums are not as significant as the one we are beginning to experience now. Dr. Zharkova, a renowned solar physicist and expert on Grand Solar Minimums, developed an algorithm looking back over 10,000 years at the occurrence of Grand Solar Minimums. She found they occurred approximately every 350 years and that means we are entering one now.

"In a video Susan discovered with an interview of Dr. Zharkova she mentioned that this one started about 2019 to 2020. Furthermore she told of discussions with farmers in Greece about how they were seeing a shortening of the growing season. By the way the video is available at our website. Tonight at 10 I will further elaborate on this Grand Solar Minimum."

"Oh my God!" exclaims Dorothy. "What are we going to do, Bob?"

"I actually heard about this Grand Solar Minimum, when an expert on Preparedness and Survival Skills was being interviewed," explains Bob. "I knew what you thought of my fascination with Bigfoot and honestly I forgot all about it until I saw the continuing drop in temperature since the first day of summer. I mentioned it to Andy and he downloaded the book the guest has written on surviving this Grand Solar Minimum.

"I saw the first thing he suggested we all have, which are BUG-OUT backpacks with numerous items in each one. Actually this is what we should have had all along with the tornadoes we see. So I asked Andy to figure out, where we get the backpacks and even a slack pack for Cindy along with the items. Fortunately the author goes into great depth on the items and where to get them. I plan to order the packs and items ASAP."

"Well this time I am so pleased you listen to that show and Andy downloaded this guest's book, which I also plan to download from Kindle," shares an enthusiastic Dorothy. "I also will share this with my sewing group."

"Mom, I am glad you are enthusiastic; however the author says the WORST thing you can do is share what you are learning OUTSIDE your family," says Andy. "I know this sounds very selfish but during something potentially as catastrophic as this impending event we need to be PREPARED OURSELVES **FIRST**. Unfortunately it sounds like and is an example of: SURVIVAL OF THE FITTEST AND KNOWLEDGEABLE."

"I see your point, Andy and as much as it seems selfish, we as a family MUST prepare ourselves FIRST," agrees Dorothy. "I do think we should not tell Cindy much about this as she could become very frightened and worst mention it to her friends."

"Could not agree more," agrees Bob as Andy affirms with his thumbs up. "I will order everything as soon as Andy gets the list together."

"Excellent," affirms Dorothy. "Right now I am going on my iPad and watch that interview."

CHAPTER 7

June 24, 2025 Topeka Kansas

Andy, Dorothy and Bob eagerly awaited the 10 o'clock weather report last night to surprisingly find there was no mention of the Grand Solar Minimum by Tom or Susan. They found this very strange.

They have tuned into the weather this morning to see why Tom and Susan did not continue their report to find a NEW meteorologist named Janice.

"Good morning and I am sure you were expecting Tom or even Susan; however they both 'resigned' effective immediately to pursue other interests," states Janice, who is reading from a teleprompter. "I am a qualified meteorologist, who will present to you the weather as it is sent to me by the US Weather Bureau and NOAA.

"So let's first clear up this speculation about the Grand Solar Minimum. Yes, there are people like Dr. Zharkova, who are definitely, who they say they are. Many of you might have watched her interview we had up for a short period of time; however our own esteemed professionals don't see any signs of this impending event. Therefore we have decided to remove it. My duty to you is to present FACT and NOT speculation. So here is what I have learned."

"Dad, how do you feel about what this new meteorologist is saying?" asks Andy.

"I honestly don't know now what to believe and think; however I still plan to purchase the items for our bug-out backpacks and Cindy's slack pack plus these sturdy backpacks and slack pack from Battle Lake Outdoors," replies Bob. "Then we should continue to monitor the temperature and IF it continues to get colder decide on a future plan. Again I think we should keep all this to ourselves."

"Could not agree more," affirms Andy.

Bob proceeds to the regular meeting of the farmers to just listen to their opinions of the developing weather and event at the TV station.

"WOW!, what do you guys make of the shakeup with our weather team of Tom and Susan" exclaims Frank. "Personally I think it is a GOOD move especially after Tom brought up all that nonsense about a Grand Solar Minimum."

The others nod their heads including Bob so he won't stand out.

Frank continues: "As Janice said this stuff about the Grand Solar Minimum is speculation by some unheard of physicist, who developed something called an algorithm, which I have no clue what that is. I certainly did not waste my time watching that video. Did any of you watch it?"

Again like bobble heads the farmers shake their heads in unison with "know-it-all" Frank again including Bob.

"We just need to follow what our government experts say," concludes Frank.

After affirmative bobbling the group discusses other matters.

CHAPTER 8

June 24, 2025 Topeka Kansas

Since Cindy has her gymnastics class and will be gone for several hours, Bob calls together Andy and Dorothy to discuss what they might have heard about the sudden resignation of Tom and Susan. Bob shares what Mister Know-It-All Frank said.

Dorothy speaks up next. "Yes, I got a call from Ethel and she wanted to talk about the resignations and ask me what I thought of it. I told her we did watch the news and weather last night; however I did not mention anything about our concerns and preparations."

Finally Andy adds what he has heard. "I heard teachers discussing it and my science teacher said he had gone up on the video link before it was removed and was very impressed with Dr. Zharkova. He told my math teacher that he was going to further research the Grand Solar Minimum and could not understand why the TV station had removed the link."

"I have been continuing to read that Preparedness and Survival Skills author's book and he says it is VITAL to have a SAFE AREA that is at least **125** miles from an urban or suburban area. We are certainly not that far from Topeka."

"So what you are saying is that we need to take out a map and find an area 125 miles from a metropolitan area," confirms Dorothy. "Also if this entire Midwest area is going to be frozen then we need to head either to the West Coast, which already is impossible in California to find a place 125 miles from already overpopulated areas. Might consider Oregon. Then south would be southern Texas and even Mexico after looking at the rendering and I really don't want to live in the drug cartel run Mexico. I see the BEST area is southern Georgia all the way into Florida."

"Yes, that is what I noted too from studying the rendering," agrees Andy. "I think we focus on southern Georgia and Florida; however Florida is close to overpopulation like California. Of course another alternative is to find a SAFE AREA, where there is an opportunity to either build underground or live in a cave."

"I can assure neither of those are options I would even consider especially with Cindy," says Dorothy. "If this event is to potentially last 35 years, there is no way I could do it."

"I certainly agree," says Bob. "At least with southern Georgia there is farmland we could purchase and basically continue the life we are living. Florida does have farmland too but as Andy mentioned it is already close to overpopulation. Therefore why don't we begin exploring what is available in southern Georgia. Again let's not alert anyone to what we are doing."

(As purchaser of this book you will have the opportunity along with nine other participants to do a Zoom call with one of the most knowledgeable land real estate brokers in this area of southern Georgia that can locate your SAFE AREA! Learn more at the end of the book.)

They all agree as Cindy comes in full of enthusiasm from her gymnastics class.

CHAPTER 9

June 25, 2025 Topeka Kansas

The weather is getting even colder even before the evening.

By the time Andy puts the livestock in the barn the temperature is 47.

"Dad, you see what the temperature is now at 6 o'clock," says Andy as he points to the thermometer. "It is BELOW what has been by several more degrees at this time. I wonder what that new meteorologist is going to have to say tonight."

"We should head on in and join Mom to watch it," agrees Bob.

As they enter the living room Janice, the new meteorologist is saying that this weather system is going to last a few more days than expected and temperatures could get even colder. However she says it is still just a freak system that is truly not the beginning of some Grand Solar Minimum as discussed by Tom.

"That woman is lying through her teeth," seethes Dorothy. "I sure hope we can get this place on the market real soon."

"Yes, I have been thinking the same," agrees Bob. "We have got to develop a 'story' of why we are selling that does not lead anyone to think we are believers in the Grand Solar Minimum."

"Well it might involve telling a 'little white lie' but we could say that we have relatives in the South, who want to sell us their farm and that we have seriously been thinking about moving," answers Dorothy. "Actually I do have some distant cousins down South; however they would not be farmers. They are so 'citified' that they probably think carrots miraculously appear in the grocery store and come from the 'carrot tree'!"

Andy and Bob join Dorothy in a good hearty laugh.

"I have been looking on the Internet at real estate websites in the south Georgia area and fortunately the acre prices are fairly decent right now but that could change soon," shares Bob. "Of course we would have to check to see if we can bring our livestock there or have to sell them here. Andy could you begin to research that and also what would be the cost of moving them there? It might be cheaper to just buy down there."

"I am going to see what comparable farms are going for currently in our area," informs Dorothy. "I know that one of my sewing ladies has a daughter, who sells farm properties as she grew up on one."

"Good, and I will be meeting with the guys tomorrow and will use your 'story' about your relatives looking to sell to us and we have been considering it not because of this Grand Solar Minimum stuff but because they are offering a really good 'don't want' price. Of course, Frank will want to know, who they are and where they live, but I will tell him that these relatives are very privacy oriented and I have been asked by them to keep their names and location out of any of our discussions."

"That is a great embellishment, Bob, and I also will follow the same way with the sewing ladies and especially the real estate agent," agrees Dorothy.

The backdoor opens as Cindy announces her return.

CHAPTER 10

June 26, 2025 Pentagon Bunker

As before all the players file into the room without aides followed finally by the Joint Chiefs Chairwoman Admiral Hutchinson.

She opens the meeting by again saying and emphasizing the importance of keeping the meeting off the radar except for key members of Congress and Senate, Vice President and President.

Her first statement concerns what she has learned in regards to the DARPA/Homeland Security project:

"Ladies and gentlemen I have learned more about this secret project and it is true that those revolutionary hand cranked generators were used by their people as they renovated abandoned mines. They are to be used to house the President, Vice President, members of Congress and Senate and other three letter agency top personnel along with their families. Yes, we are all on the list."

She continues:

"Already a hedge funder has bought abandoned US missile silos and turned them into luxury condos and they are all SOLD OUT starting at **$5,000,000** and going up. Also our Pentagon intelligence has confirmed that Vladimir Putin has taken one of their underground Space War facilities and turned it into condos for himself and his multi-billionaire thug buddies.

"As you can certainly ascertain the President has set aside areas for a number of multi-billionaires like Elon, Warren, Jeff and others he feels are vital to the future of this country."

Snickers are heard through out the bunker room.

"Whether we agree or not he is still our Commander-In-Chief and his wishes are our marching orders."

"So Madame Chairwoman, where does that leave my wonderful next door neighbors," says General Davidson of the Marine Corps. "Are they to be left to fend for themselves against both freezing Arctic temperatures and starvation?"

There are murmurs of agreement.

"General Davidson and every person in this room I share your same concerns; however currently the priority is securing the aforementioned personnel and ourselves," states the Chairwoman. "I will say I have actually read that author's book since we met and he offers some excellent scenarios for survival for the others. In fact, his mission through this book is to make as many people aware of the current Grand Solar Minimum and its effects. He actually has put together what he calls a Braintrust based on the ones he learned about at both Pixar and Disney Animations."

"That's a cool name," says General Smith of the US Army.

"Actually General Smith you might remember that decades ago one of your colonels wrote this book I am holding in my hand that totally fascinated me and gave me hope that our US Army and eventually entire US Armed Forces might see a new vision of what they could become," declares the Madame Chairwoman Admiral Hutchinson as she holds it up for everyone to see.

FIRST EARTH BATTALION OPERATIONS MANUAL

Evolutionary Tactics

REPRINT OF ORIGINAL MANUAL FROM THE 70'S

JIM CHANNON

"It seems this author also read the book by Colonel Jim Channon, where he calls for the First Earth Battalion under the direction of the US Army," continues the Chairwoman looking directly at General Smith. "One particular area under this First Earth Battalion was to be called The Natural Disaster Rescue Group and if this had been put in place then we would have had a mechanism in place to deal with this Grand Solar Minimum.

"Also he called for The Human Disaster Rescue Group, which now would be, who we would be listening too instead of just beginning this late in the game to figure our next steps. So now a regular citizen has taken on the burden of following the suggestions of this forward thinking US Army Colonel, whose brilliant strategies and tactics mentioned through out the manual were deep sixed by bureaucratic higher-ups."

"As I remember Admiral Hutchinson, this founder of Earth Battalion is the same person, who was mentioned by NOOA meteorologist Dr. Edwards," explains General Jackson of the US National Guard. "Dr. Edwards read in this author's article in I think **Survivors Edge** that there were numerous ways to help people survive – caves, subway stations, basements – any place **10 FEET** or more underground. I also have read his book and one ingenious scenario is using mothballed AIRCRAFT CARRIERS to house potentially thousands on each ship. So what do you all think of this scenario?"

"I also read that General Jackson and like you I think it is ingenious and more importantly doable," agrees Admiral Hutchinson. "Now I would like to defer to Admiral Stetson."

"Unfortunately I have not had the opportunity to read what you and General Jackson have; however I just downloaded it on my phone," replies Admiral Stetson. "What I am hearing is truly ingenious and doable as you have mentioned.

"I would have to connect with the ones in charge of the mothballed aircraft carriers and see their current condition. If they are in good enough condition to put back in operation, then I will have to develop a reason for doing it so we don't panic the potential civilians refitting them."

"Excellent point, Admiral," agrees Admiral Hutchinson. "What if we put out a story that one of our allies especially less wealthy than us is interested in our mothballed carriers for their own defense. Could that fly? "

"That is certainly a possibility if we can get that ally onboard with this story without setting off alarms to the media," replies Admiral Stetson. "Any suggestions from this room are certainly appreciated."

"Now on to another matter brought to my attention by the other NOAA meteorologist Dr. Matheson," says Admiral Hutchinson. "It seems a meteorologist in Topeka, Kansas decided to not follow the suggestions of keeping the general public in the dark about the Grand Solar Minimum. He learned from his fellow colleague about it and Dr. Zharkova and more importantly her interview.

"He watched it and began to gather more info from the Net and decided to go public with it on his 6 o'clock weather segment and tell the viewers how they could watch Dr. Zharkova's interview too from the link he provided at the station's website.

"NOAA heard about it and got in touch with the top management of the broadcast company that owned the station. They told these honchos to get those two off the air ASAP if they planned to still be in business! It was quietly arranged with the so called resignations of both of them, which the top honchos said for them to go quietly into the night with their new hush up money, which was substantial.

"A new by-the-book NOAA meteorologist has taken their place. She immediately shot down the 'speculations' of the two and went by the book with the 'weird weather system' story."

"Thank goodness they stopped this from going any further, says a relieved General Jackson. "I just pray the people, who watch this channel believe the new meteorologist and not the information the others shared. Is it my understanding that the link to Dr. Zharkova's interview has been removed?"

"Yes, it has, general," affirms Admiral Hutchinson.

As there is no further business the Joint Chiefs adjourn to formulate how they are going to "spin" the ally story concerning the mothballed aircraft carriers.

CHAPTER 11

June 26, 2025 Topeka, Kansas

"Dad, that Preparedness and Survival Skills consultant we have been studying through his book is going to be on your favorite show," says Andy as they are studying the continuing drop in temperature at 6 o'clock on a summer June day.

"Yes, and I can't wait to hear him," replies an excited Bob. "I plan to ask Mom to listen too along with you and I."

"Great as I can't wait to hear him too," continues Andy as they head to the house.

Sure enough Dorothy is excited to hear the show especially since it is on a topic she wants to learn more about – Surviving The Grand Solar Minimum.

The host George introduces the audience to the guest:

"Tonight I am pleased to have back our guest talking about surviving the Grand Solar Minimum. He like myself is a former investigative reporter, who first came to my attention when he appeared as the first one to discuss the implications of the past potentially catastrophic event - Y2K – and how to survive it. Welcome!"

"Thanks, George," continues the consultant . "And yes tonight we are going to discuss the ways to SURVIVE this potential 35 years event of Arctic freezing temperatures and the ever shortening growing season."

"First let's say hello to renowned solar physicist Dr. Zharkova, whom you brought to our attention in your last appearance," interrupts George. "Welcome Dr. Zharkova."

"Thank you gentlemen," replies Dr. Zharkova. "It is an honor to be on the show."

"It is our honor and good fortune to have you here, Dr. Zharkova," replies George. "Dr. Zharkova it is my understanding that you created an algorithm that looked at the occurrence of previous Grand Solar Minimums over a 10,000 year period. Also you had a colleague take this same algorithm and she got the same results as you did."

"That is correct George," states Dr. Zharkova. "So I advanced it 350 years ahead and correlated with the observations we were seeing currently. Everything lined up especially the colder weather that we are seeing. I decided back in 2019 to share this information on YouTube and if I am not mistaken your other guest watched that video that I shared with fellow physicists in the UK, where I have continued my research after leaving the former Soviet Union."

George asks Dr. Zharkova to explain what occurs to create a Grand Solar Minimum, which she does.

"Thank you Dr. Zharkova for this explanation," says George. "Please stay with us as long as possible. We will be right back after this break to take your calls for Dr. Zharkova."

After the break the phone lines are jammed with callers and Dr. Zharkova spends the rest of her allotted time answering the questions being asked.

"I know it is late in the UK and again I thank you for staying with us as long as you have," says George. "We will have your http://www.solargsm.com up for all you listeners and others, who are regular subscribers. After the break we will be back with my other guest and former news reporter colleague."

"I wonder what he is going to say to night?" contemplates Bob.

"We are back and are you in for a avalanche of information for surviving this Grand Solar Minimum," starts George. "So take notes and at the end of his segment we will have his website available like we have done with Dr. Zharkova. Take it away my friend!"

"Thanks, George and as always a pleasure talking with a REAL newsperson as opposed to a 'pretty blue eyes bobbling head reading from a teleprompter' as Captain Courageous Ted Turner referred to current anchors on the news shows," quips his friend.

"First, let's realize that what we are facing now makes Y2K look like Romper Room. Even the previous COVID outbreak does not hold a candle to the ramifications of potentially 83,000,000+ Americans in the cross hairs of this epic catastrophic event with Arctic freezing temperatures, shortened growing seasons and starvation. As you remember from my previous time on your show the last Grand Solar Minimum created a Mini Ice Age and 25% of the European population DIED of STARVATION! Heck the American Revolution had not occurred then!

"During and after COVID we saw the destruction of vital SUPPLY CHAINS leading to one of the biggest recessions and in my opinion bordering on a DEPRESSION.

"As you remember our Federal government did NOTHING constructive to bring us out of COVID except throw TRILLIONS of stimulus money at it leading to the huge recession and no solutions to the so called 'epidemic'.

"We are seeing the same ineptness with handling the Grand Solar Minimum that the great Florence Nightingale saw during the Crimean War. Let me give the audience some very appropriate quotes from this nurse and more importantly statistician:

'Foul air and PREVENTABLE (my emphasis) mischiefs…gangrene, lice, bugs and fleas…no mops, no plates, no wooden trays, no slippers…no knives and forks, no scissors for cutting the men's hair, which is literally alive…no basins, no towelling, no chloride of lime.'

"She is referring to when she arrived at the Barrack Hospital at Scutari, opposite Constantinople, Turkey. She saw: 'four MILES (my emphasis) of corridors of wounded men sleeping 18 INCHES (my emphasis) apart from each other on thin pieces of cloth'!

"So why did these horrible conditions exist?" asks George.

"A broken SUPPLY CHAIN and incompetent BEAUROCRATS just like we are seeing today in 2025 as she saw in 1855!" replied the guest. "She ON HER OWN found the necessary supplies around the area. She took matters into her own hands and YOU better do the same! We think we can solve everything through AI – Artificial Intelligence – however what everyone seems to forget is AI happens because of SUPER computers.

"And WHEN NOT IF the GRID GOES DOWN what good is AI, cryptocurrencies, wearables, smartphones, tablets, laptops, ATMs, etc. with NO electricity? Again you better have HAND cranked items like I have and a HUGE supply (at least a pallet) of BATTERIES.

"So do you think that the Federal government heads including the President are as unprepared as over 90% of Americans?" wonders George.

"Think again!" fires back the consultant. "George, you know those revolutionary hand cranked generators I have had at my website, which currently are NOT available due to the SUPPLY CHAIN problems, actually were used for three years in a DARPA/Homeland Security project NOT on US soil but in MOROCCO!"

"What!" exclaims George along with Bob, Dorothy and Andy.

"Yep, Morocco where they dug out the desert to put in UNDERGROUND living quarters built using ancient building principles," explains the consultant. The generators provided the necessary power.

"Furthermore the US and UK provided the funds to build a $650,000,000 WINDFARM in Morocco. It is now reported that the sea harbor of Talfaya, Morocco has been extensively deepened to take in the super rich's mega yachts. Both Bill Gates and the Clinton's are reported to have new digs there.

"So you ask why has Morocco been the chosen safe place?

"Real simple. Already US wealthy families are sending their kids to the Moroccan universities like this one you will see at my website outside of Guelmim known as The Technology School. It is built following the latest architectural trends in Eco Design and Eco Development allowing for this minimalistic facility to be plopped down in the desert.

"Another reason is another potentially catastrophic event that occurs every 2000 years or so when the Atlantic Ocean side massive wall on the Canary Island falls into the ocean causing a tsunami that when it reaches the USA will have a wall of water **150 FEET** high from Jacksonville, Florida to Boston. This wall could collapse into the

Atlantic any time and the volcano on the island does not help matters due to its recent eruptions.

"They believe this wall of water will sweep over Florida continuing to build further in the Gulf of Mexico. Then it will sweep across basically flat land all the way to the Pacific. Then it will again build itself up until it goes across the Pacific and amazingly back towards the same Canary Island.

"For some reason these super wealthy believe Morocco will be a safe place from it and the Grand Solar Minimum. Of course these IDIOTS have not considered the upcoming 2030 MOON WOBBLE event that will FLOOD coast lines around the GLOBE and for me Phase 2 of what I am calling **GOD'S RESET**.

"No wonder our producer has named you Doctor Gloom," says George. "So where will anyone be SAFE from these events?"

"Ironically one of the best places could be MEXICO!" laughs the guest. "The USA has been building a wall to keep out the illegals. During especially the Grand Solar Minimum we could have literally MILLIONS of Americans trying to cross over into Mexico! In fact, a recent main stream media story confirmed this exodus! However as more flee, I bet they will be met by the Mexican authorities and turned back!

"Another potential safe area could be the Caribbean – Puerto Rico, US Virgin Islands, British Virgin Islands, Antiqua, etc. My good friend and real estate broker on St. Croix US Virgin Islands has told me that the CHINESE and Saudis have been snapping up every available estate, condo and especially ocean front property. I hear it is also happening in Australia and New Zealand, both places I considered relocating to.

"Central America especially BELIZE is a great possibility as I have an excellent connection there even though other areas of Central

America and northern South America are going to see adverse effects from the Grand Solar Minimum.

"I tell our world wide clients to be MOBILE. With a sailboat or cabin cruiser you have that mobility. I plan to go that route myself.

"Yes, the Canary Island tsunami will force me to head way south along with the Moon Wobble; however at least I will have had my Coast Guard training and possess my Captain's license. Instead of location, location, location I say MOBILITY, MOBILITY, MOBILITY."

"WOW!" exclaims George. "After the break we will be back to answer your questions."

After the break, the phone lines are lit up like a Christmas tree and require overtime to answer them cutting into George's next guest's time.

"Always good to have you, my friend and plenty for all of us to think about," says George.

"More than happy to help your listeners," replies the consultant.

"So with what you have heard, Dad, do you think south Georgia is still good for us or should we consider further south into Florida and live on a boat?" questions Andy with Dorothy looking at Bob.

"After what I have heard tonight I haven't a clue," shares a perplexed Bob. "I know one thing. I totally agree that these events coming are truly GOD'S RESET. I think the best thing we can do is go and pray and more importantly LISTEN!"

"Amen," says both Andy and his Mom.

CHAPTER 12

June 27, 2025 Pentagon Bunker

An emergency meeting has been called by the Chairwoman of the Joints Chiefs Admiral Hutchinson.

"As some of you know that internationally syndicated radio show had both Dr. Zharkova and the Preparedness and Survival Skills consultant on it last night," begins the Chairwoman. "Already their huge listening audience of up to 5,000,000 are calling their members of Congress, Senators, governors and mayors DEMANDING answers to what they plan to do for OUR citizens and NOT Moroccans.

"Yes, that so called secret project is no longer a secret. They are visiting both Dr. Zharkova's and the consultant's websites to see that rendering that we have seen to see if they are one of the 83,000,000+ Americans in the cross hairs of the current Grand Solar Minimum that the consultant has spoken about."

Groans are heard through out the Bunker.

"Frankly trying to do a 'spin' story of disinformation is no longer an option," continues Admiral Hutchinson. "And the President agrees and wants a plan he can present to these panicked Americans with substance."

More groans are heard.

"So let's stop moaning and groaning right now and do as the President has ORDERED," commands the Chairwoman. "Since you are in charge of the National Guard, General Jackson, you will head up the future meetings. I know that the scientists that met with us before said the consultant had good ideas about using the underground subway stations, huge basements of buildings and even the 205,000 square miles of cave systems across the United States."

"Yes, I took extensive notes on those proposals," states General Jackson. "I then began to research their viability and have to agree that all three offer viable solutions for a LIMITED amount of Americans. Unfortunately I don't have the numbers as that would have required a 'need to know' from the experts."

"General you have my permission to get those experts crunching the numbers ASAP," replies Admiral Hutchinson. "I am sure the President, Congress and the Senate are going to need those numbers as reporters start to bombard them with questions."

"Thank you Admiral," says General Jackson. "I am going to need potential solutions from the Corps of Engineers, which comes under the command of General Smith."

"You have my full cooperation General," replies General Smith. I think these experts can best be used to generate ideas and solutions for reinforcing these 205,000 square miles of caves just like was done with those abandoned mines. We may again need that supplier of those hand cranked generators that were used in the Moroccan project. Yes, one of my aides brought to my attention the show and replayed it for me.

"Furthermore General Jackson I will be checking in with my general at Fort Campbell in Kentucky to see how we can build up as many troops as possible to handle the potential chaos and rioting along with your National Guard."

"Thank you General Smith," says a relieved General Jackson.

"Admiral Stetson, what is the feasibility of re-commissioning those mothballed aircraft carriers that the consultant mentioned?" questions the Chairwoman.

"Actually I too have heard a replay before this meeting as you remember my aide listens regularly," states Admiral Stetson. "It is certainly feasible though a mammoth task. I can begin discussions with the admiral in charge of the mothballed aircraft carriers."

"Excellent," replies Admiral Hutchinson.

"So what is this old leather neck suppose to do? " asks General Stetson.

"Don't worry general I have plans for you," replies back the Chairwoman. "There is going to be potentially dangerous scenarios develop at our world wide embassies as knowledge of the ramifications of the Grand Solar Minimum spread. I want you to present to us a plan to protect these embassy personnel and their families using your Marines stationed at the embassies."

"Consider it done, Admiral Hutchinson," says General Stetson. "I will further prepare our Marine Recon personnel to be ready to assist."

"Okay, we all have our assignments so let's not waste one more minute," states the Chairwoman. "Don't answer any questions from the press sharks that will be hovering over the Pentagon more so than ever. Refer them to my office and we are for now adjourned."

There are a number of discussions that continue in an atmosphere of hope and excitement.

CHAPTER 13

June 27, 2025 Topeka Kansas

All the family less Cindy, who is off visiting a friend, are seated around the kitchen table enjoying a second cup of coffee.

"Okay, we need to pray for guidance in this meeting we are about to have," begins Bob.

They hold hands together as Bob leads them in prayer.

"Andy, you are the one, who knows probably more about the options we learned of on the show, so lead us," continues Bob.

"Sure, Dad," begins an enthusiastic Andy. "First, we have the south Georgia option of starting another farm. The rendering showing the previous coverage of the Mini Ice Age over the not even created United States, shows that the Jesup, Georgia area was spared. Dr. Zharkova has stated we should not see a repeat of the Mini Ice Age; however the consultant spoke of the Canary Island tsunami that is way past due. It is possible that Jesup, Georgia could be flooded; however it is 95 miles from the Atlantic Ocean.

"However there is a potential solution to it and other weather problems like the Moon Wobble. I have been reading in the consultant's book about his DODECAHEDRON Raptor Domes Kits™ that are fascinating.

"First some history of the dome. It is by far the STRONGEST man-made structures and used through out history. We even have one on top of the US Capitol. As you know it has withstood centuries of weather.

"Along came Buckminster 'Bucky' Fuller and his fascination with the geodesic dome even though he did not create it but became the evangelist for its use in the USA. It received some use but never caught on as it should have.

"MIT or the prestigious Massachusetts Institute of Technology decided to improve on the geodesic dome resulting in the dodecahedron geometric shape used by the consultant, who incidentally studied with the right hand man to Bucky.

"This shape is STRONGER and is able to withstand Category 5 hurricane winds and important to us…waves. It is built with 2x6 southern pine struts. This is again good for us as these southern pines are through out the Jesup area.

"We could actually mill our own struts. Then there are the plywood hubs that connect with even wooden dowels or bolts and nuts. This makes up the superstructure and sits on a concrete foundation of between 3 and 6 inches. Dad and I have laid concrete before.

"If we place the dome or domes on a higher set of rebar re-enforced concrete pillars like they have done on the beaches of Florida, then the tsunami waves will go around and over the domes. Ditto for hurricane waves. The Moon Wobble waves will not likely reach Jesup as it is about 95 miles from the Atlantic Ocean as I mentioned."

"What are these domes covered with, Andy?" questions an excited Bob.

"Dad, there are several options for covering them. One is to nail on triangular pieces of ¾ inch marine plywood that covers the surface of each triangular space. Then GAF 'environmentally friendly' HydroStop® is brushed or sprayed on. This covering should last as long as ten years before it needs re-coating.

"Another option that really excites me is using HempCrete to fill in the triangular spaces. First it is easy to make using the outer husks of hemp that is abandoned by the makers of CBD products that are sold everywhere and on the Net. You then mix these shredded husks with water and lime. Second it can be trialed into the spaces or sprayed into them. Third HempCrete will outlast the HydroStop® by decades. And fourth it acts as an insect repellent. We could easily learn to make HempCrete.

HEMPCRETE
DESIGNED TO BUILD, NOT TO SMOKE.

- HEMP + LIME + WATER
- ENERGY EFFICIENT
- LASTS 100'S OF YEARS
- FLAME, WATER & PEST RESISTANT
- NATURALLY NON-TOXIC
- INCREDIBLE INSULATION
- STRONG, LIGHTWEIGHT & BREATHABLE

TheFreeThoughtProject.Com | #EndTheDrugWar

BEST PART- ONLY TAKES 90 DAYS TO GROW!

"The dome struts and hubs can be cut on a CNC router programmed with the numbers of the dimensions of the struts and hubs. We should easily find a shop with a CNC router that would gladly work with us. Or the consultant has access to a Chinese CNC router manufacturer that has assured him that their router can cut the struts and hubs. It is priced at under $10,000.

"We could sign a royalty agreement with the consultant to pay him his requested royalties for the four or five sizes he has the dimensions for both the struts and hubs. I like this option best because then we can have another business to fall back on if the farm is wiped out by the tsunami, a hurricane or drought like they are experiencing out West."

"WOW, Andy kudos for a great presentation on this Jesup option," congratulates Bob along with Dorothy clapping enthusiastically.

"If Mom can now have her say as ladies first," teases Dorothy. "I like this dome idea and I also like the dome manufacturing business. Frankly I would rather do this than run a commercial farm again.

"I see great potential in these domes for surviving the Grand Solar Minimum, the flooding of coastal areas with the coming Moon Wobble and just as a place to feel safer due to its geometrical structure especially this dodecahedron one."

"Gosh, Dorothy I am shocked and pleased at what you have said and agree 100%!" replies Bob.

"I am also pleased, Mom, you like the domes and the potential of manufacturing them," enthused Andy. "I forgot to mention that these Raptor Domes Kits™ withstand also T4 tornadoes that we have experienced and they experience in the Jesup area and Richter mid-level earthquakes, which are less likely in our area or Jesup," says Andy.

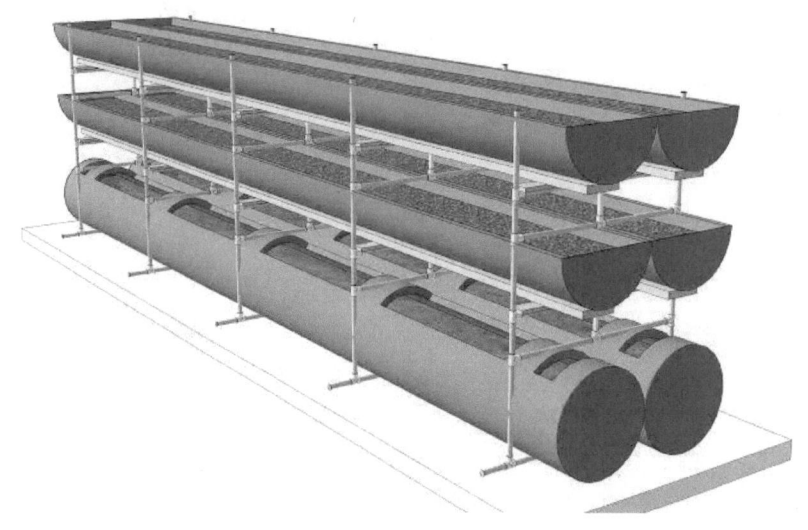

"Also a physician has developed a revolutionary water-based growing system in three troughs, where the first trough raises fish or crayfish and their nitrogen waste is fed to the other two troughs, where veggies, fruits, wheatgrass, herbs, rice, etc. can be raised," continues Andy. He has married this with the Raptor Dome Kit™ and the consultant offers Raptor Bio-Domes™. The power needed to run the system can be generated by solar panels."

"Another possibility for the dome kits, that I personally could really get excited about!" says Dorothy.

"I suggest Andy contact the consultant and we sign this royalties agreement if we can afford it," suggests Bob.

"You both are going to love this," replies Andy. He is also a Christian and sees the dome manufacturing business as a ministry especially for low income housing.

"His blog states that a turnkey ready to move in Raptor Dome™ of 32 feet in diameter offering 805 square feet on the first floor with an optional loft of up to 200 square feet can be done for around **$75,000**! That is cheaper than most tiny homes and double wide modular homes.

"And the royalties agreement requires NO UPFRONT FEES only the royalties, which still leave a potential NET profit of **30 – 60%** depending on the size!"

"Get in touch with him ASAP, Andy!" says Dorothy and Bob simultaneously.

CHAPTER 14

June 28, 2025 Topeka Kansas

Bob heads out to his morning meeting with the guys full of excitement as to the new plans of exiting Topeka. Of course, he has no intention of sharing the destination with the group but plans to put out "bait" with especially Mister Know-It-All to see if there might be interest in selling the farm to him. Bob has a gut level feeling that Frank will still not believe in the Grand Solar Minimum.

"Well guys what are we going to discuss today?" questions Bob.

"I have seen this morning on the Web numerous discussions after that radio show about this Grand Solar Minimum," offers Ed.

"Not that Grand Solar Minimum crap again," moans Frank. "Who was discussing it? I bet it was that looney so called physicist that those 'fired' meteorologists talked about."

"Actually you are right, Frank," confirms Ed. "Personally I like you think it is crap. What about you Bob?"

"I am afraid I have not heard the show and have no opinion currently," says Bob.

"Hey, Bob, you still considering that move to help Dorothy's relatives?" questions Frank.

"Yes, Frank, answers Bob. "In fact Dorothy is going to be talking to a real estate agent, who is the daughter of one of Dorothy's sewing group."

"WOW! so soon the group will be losing you?" exclaims Frank. "Hey I might be interested in your property myself so give me the opportunity to make an offer."

("Got you Frank – hook, line and sinker!" thinks Bob. "But I will still play it cautiously.")

"Sure, Frank," replies Bob. "After the real estate agent and an appraiser give us their results you will certainly be told."

"That is all I ask, Bob," replies Frank. "So what else can we discuss?"

The discussion goes on another hour and then the farmers head for morning chores.

Bob can't wait to get home and tell Dorothy that as he predicted Frank is interested!

CHAPTER 15

June 28, 2025 Topeka Kansas

"So Mister Know-It-All played right into your game plan," says Dorothy.

"Yep, he took the bait like the hungry shark I consider him," laughs Bob. "He has been wanting our farm for as long as I have been in the group. Every now and then he has hinted to me what he said out right this morning."

"I will get with that real estate agent ASAP as her mother gave me her number," says Dorothy. "I am sure she will also know a honest experienced appraiser."

"Good plan," affirms Bob as he heads out the door to his morning chores.

Dorothy immediately calls Jane, who is the real estate agent.

"Hi, Jane," begins Dorothy. Your mother is in my sewing club and told me about you being a real estate agent and gave me your number. I am Dorothy."

"Yes, my Mom told me she had given you my name and number," replies Jane. "I would love to work with you as your seller agent. In today's real estate market there are specialties and mine is as a seller agent and I am also a broker.

"I further specialize in farms as I grew up on ours learning every aspect of farming. I was very active in our 4H club. After graduating high school I went to Kansas State University and studied agriculture and farming. I also studied business and marketing and decided to pursue becoming a real estate agent and broker specializing in farms. I have also been mentored by Grant Cardone, whom **Forbes** considers the Number 1 sales trainer. He has been a long time consultant to Century 21, where I am an agent/broker."

"WOW! I truly know GOD has led me to the right person," exclaims Dorothy. "When can we get together as we would like to move ahead as soon as possible? My cousin in Georgia needs for us to come there and take over their farm."

"Would this afternoon at 1:45 be good for you?" questions Jane. "I am booked till then."

"Perfect, and my husband Bob will be available too plus it will give me enough time to straighten up the house'" replies Dorothy. "I would like to finish before Cindy, my youngest, gets home at about 4 as she doesn't know anything about the potential move."

"I will see you promptly at 1:45 along with your husband Bob as both of you will have to sign the selling agreement if you desire to move ahead," says Jane. "I will be finished by no later than 3 as I have another appointment at 3:30."

Dorothy heads out to the barn, where she finds Bob re-shoeing one of the horses. She informs him of her conversation with Jane.

"She does seem like our answer to prayer," says Bob. "I definitely look forward to meeting her."

"Me too," answers Dorothy. "Now I'm going into the house and tidy it up."

At exactly 1:45 Jane drives up and after the greetings in the front yard, they head towards the front door.

Jane stops to look at the front porch with the flower boxes in each front window along with the swings at both ends of the porch.

They enter the front door heading into the modern open floor plan of living, dining and kitchen.

"WOW! I was definitely not expecting this!" exclaims Jane. "Today even farm families or wannabes look for the open floor plan. I also note though it is open floor plan you have kept the farmhouse fireplace, which on an usually chilly day like this is very inviting."

"I am glad you are pleased with my open floor plan and other things we did upon purchasing the farm," replies Dorothy. "I may be a farm girl but I watch HGTV and read the latest interior design magazines even having a subscription to **Architectural Digest**, especially enjoying their online **Clever** magazine."

"I can definitely take some lessons from you," replies Jane. "Let's see the rest of the house.

As with the open floor plan Jane notes the other two rooms that sell houses – bathrooms. One features the old fashion claw feet tub that Dorothy found one Saturday. In addition there is a glass enclosed shower with rain forest shower head. The second bathroom for the ensuite is more modern with the beautiful deep oval tub and touch taps. It has the same glass enclosed shower with rain forest shower head.

"Your bathrooms definitely pass with flying colors," says Jane. "On to the bedrooms."

Jane finds four very good size bedrooms especially the master, which is furnished with traditional farmhouse furniture and a quilt Dorothy has made.

Jane also discovers a modern laundry room with a sink and modern washer and dryer. She also notes a sewing machine and chair.

"Yes, this room doubles as my she cave," laughs Dorothy. I love to sew to the peaceful sounds of both the washing machine and dryer."

"I seldom encounter a house that is ready to be sold," says Jane. "Yours ticks off all the boxes including the back porch I am now seeing. What type windows do you have? How about the HVAC system?"

"This is my time to hopefully shine and tick off more boxes," answers Bob. "The windows are all triple paned and the HVAC system was just updated last year."

"Definitely more boxes ticked off as I had expected," replies Jane. "If the barn is in as good condition as this house, then we could sell this farm in less than a month!"

"WOW! We did not expect it could sell that soon!" exclaims Dorothy followed by Bob.

"Absolutely," confirms Jane. "Let's tour the barn as our time is getting limited. You both still have to sign the contract if you desire to use me."

"Let's do that right now if Dorothy agrees," answers Bob. "Done deal," confirms Dorothy.

They proceed to sign the seller's contract for a period of six months.

Next they head for the barn, where Bob answers all of Jane's questions.

Jane shakes hands with her new clients and drives off.

"I think it is time to tell Cindy the plan so we can spend time explaining why we are moving but ask her to keep it a secret as little girls like her like to do," shares Dorothy. "I will make it a game with her."

"Great idea!" exclaims Bob.

Just as Bob is returning to his work in the barn the UPS truck pulls up.

"I bet he has the backpacks and slack pack for Cindy," says Bob.

Sure enough after unpacking the boxes it is the three backpacks and slack pack ordered from Battle Lake Outdoors.

"This will make the game I plan to play with Cindy even more fun," says Dorothy.

Just about then Cindy walks through the door all excited about her time in gymnastics, which she shares with her Mom and Dad. She also notices the packages on the table.

"What is in those packages?" asks an excited Cindy. "Is there one for me?"

"There certainly is," says Dorothy handing Cindy her package.

Cindy is thrilled with this discovery and tries to open it herself but needs the help of her Mom.

"WOW! How cool is this!" exclaims Cindy. "My very own slack pack and much stronger than those flimsy ones other kids have!"

"Yes, and this one is not only for your school books and things but also what is called a BUG-OUT slack pack," explains Dorothy.

"A bug-out slack pack," says a confused Cindy. "What does it do? Keep bugs out?"

"Not exactly; however that could happen if we made up a mixture of natural ingredients and put it all over the slack pack," says her Mom. "Actually it is used in case of an emergency like you practice for with the tornado drills in school.

"It will contain all the items you will need if we have to leave the house because of a tornado heading our way. Your brother, Dad and I will have larger heavier backpacks with more items we might need during the time we are gone from the house."

"I now understand why it is called a 'bug-out' slack pack because we might need to BUG-OUT!" laughs Cindy. "How cool is that! I bet NONE of my friends have one! So what can I put in mine?"

"You can put in several of your smaller plush toys, a video game or two, some clothes, a pair of sneakers with several pair of socks," answers Dorothy. Also several books including your Bible, photos of the family and friends, packages of AA and AAA batteries, and possibly some other items. Definitely the charger for your phone."

"AWESOME!" exclaims Cindy. "I want to take it to my room and start loading it up!"

"Sounds like a plan and then you can show all of us after dinner," replies a relieved Dorothy.

Cindy heads with new slack pack to her room.

"That went way better than I expected," says Dorothy to Bob, who gives a thumbs up. "Now I pray the impending announcement of our move goes over as well. Let's you and I pray right now."

Dorothy and Bob pray for the proper guidance to tell Cindy of the upcoming move.

Bob returns to finish chores and share what has taken place with Andy.

It is after dinner and Cindy is excitingly showing and telling her family about the current items in her new slack pack.

"Cindy we have some more to share with you; however for now it is our secret to be shared with no one," explains her Mom. "You are always telling me you like to have secrets even from your best friends."

"Yes, so you have a NEW secret for me?" questions an excited Cindy.

"Yes, and it stays within our family for at least the time being," further explains Dorothy.

"You can count on me NOT to share it!" exclaims Cindy. "So what is it?"

"Okay, there is a weather event coming soon that could be very hard for all of us to deal with," starts her Mom. "In school you have studied the Eskimos that live in the very cold North Pole area. You told me though their clothes were cool, that it was way too cold to be there."

"Yes, I would never want to live there," says Cindy with strong conviction. "We are not moving there are we?"

"Not as far as I am concerned," says her Mom and there is total agreement with her Dad and Andy.

"Unfortunately this weather event could bring the very conditions of the North Pole to Topeka," explains her Mom. "It happens every 350 years or so and is called the Grand Solar Minimum. I am sure you have felt how cold it is already on these past summer days."

"Yes, my classmates in summer school have been talking about it too and several mentioned what you just said," states Cindy.

"So what do they think about these cold summer days?" pipes in Andy.

"They don't like it and I don't either," replies Cindy. "The Fourth of July is coming and I want to go swimming like we always do before the evening picnic and fireworks. If it keeps getting colder, I won't get to do it. Plus after summer school won't I be going to camp again? At this rate the lake could be like the North Pole!"

"So here is the secret," says Dorothy. "We are going to put up the farm for sale soon and move to, where you just may be able to swim year around as Georgia is way warmer than Kansas though I guess there will be some colder times. Plus the area we are looking at is less than a hundred miles from the Atlantic Ocean."

Everyone waits to see Cindy's reaction.

"COOL! Awesome!" exclaims a very excited Cindy. "When do we leave? Can we do a road trip before the Fourth of July? I just hope there is a gymnastics class at the school there. What a COOL secret!"

"You are not upset about moving away from your friends especially your sleep over buddies?" questions her astonished Dad.

"Not really as they can do a road trip to see us in the future," explains Cindy. "You know how I like ADVENTURES and this move will definitely be one! Plus I will be making new friends!"

"You certainly will!" says Andy. "I will be doing it too!"

"Remember for now this is OUR secret," reminds Dorothy to the family.

"Yes it is!" exclaims Dad, Andy and Cindy.

Soon Cindy is off to her room to begin preparing for the move. Dorothy, Bob and Andy soon follow after a prayer of thanks!

CHAPTER 16

June 30, 2025 Topeka Kansas

"I am still amazed how Cindy responded to the impending move, Jane," as Dorothy relates it to her new real estate broker. So let's get this show on the road as we are going to surprise Cindy with her 'road trip' idea to see our potential new place to live over the upcoming Fourth of July weekend."

"Smart and yes, I can call Larry, the real estate appraiser I mentioned and see his earliest availability," says Jane. "Also I will call my videographer buddy Jack to shoot a selling video plus fly his drone over your property."

"As Cindy would say: AWESOME!" laughs Dorothy.

"By the way, by using Century 21, I will be able to put up your farm across the USA and around the globe," shares Jane.

"Now I see why God pointed us to you," replies Dorothy.

Dorothy decides to look up on the Net Jesup, Georgia to see if there is a motel that still can take them for the Fourth of July weekend.

She finally locates one and books it.

Later at the dining table Dorothy decides it is time.

"Cindy, I have another surprise for you," teases her Mom.

"What is it?" exclaims Cindy. "Don't hold me in suspense!"

"Remember when you suggested a 'road trip' to Jesup, Georgia during the Fourth of July weekend," teases Dorothy again.

"On my goodness!" exclaims Cindy. "Are we going?"

"Yes!" exclaim Dorothy, Bob and Andy on cue.

"Thank you, thank you!" exclaims Cindy. "I have to go right to my room and get everything ready!"

"Yes, and don't forget to include your bug-out slack pack," adds her Dad. "Excellent opportunity to have it ready as we also will have ours packed too."

CHAPTER 17

July 1, 2025 Topeka Kansas

"Road trip today! " chimes in Cindy as she comes to the breakfast table, where Bob, Dorothy and Andy are already seated. "I have everything ready including my slack pack."

"Good for you Cindy," agrees her Mom. "And about you two gentlemen?"

"Done," says Andy.

"Ditto, for me," says their Dad.

It is a blessing to have Ed watch over everything for us," agrees Dorothy.

Several hours later the family pulls out of the front of the house.

"The temperature is calling for POOL time at our motel!" exclaims Cindy as she looks at her smartphone.

"I have also read they are having an AWESOME Fourth of July celebration with the biggest best fireworks in Jessup's recent history," states Andy.

"How COOL is that!" exclaims Cindy.

"We are also stopping over in Franklin, North Carolina to meet with the Preparedness and Survival Skills consultant that your Mom, Andy and I have heard on Coast to Coast AM and read his book," explains Bob.

"What is there to do there?" asks Cindy.

"Cindy, Franklin is known as the Gem Capital and they have places you can go, where you pan for these gems," says Andy.

"WOW, how COOL is that!" exclaims Cindy. "That would be AWESOME to do! Can we do that while we are there?"

"Sure," says her Dad.

As they approach Franklin from the west going east they begin to see the incredible mountains that surrounds them.

"Look at these mountains!" exclaims Cindy. "They are like nothing I have ever seen out in Kansas!

"You are right about that, Cindy," agrees Andy. "We are lucky to be able to see these mountains and as far as the eye can see."

Bob and Dorothy are also awed by the majestic Smokies. They like Cindy and Andy have never seen mountains like these.

Within an hour they are sitting in a restaurant in Franklin next to a gas station waiting for their new friend to arrive.

Just then in walks the Preparedness and Survival Skills consultant, who is fit, trim and tanned with salt pepper hair. He is dressed in shorts and a Raptor Domes™ tee shirt.

"Sorry the bus was running late," he states as he greets each of them by name. "My apartment complex supplies free monthly bus passes and the busses will take me wherever I want to go."

"So you don't live in one of your own Raptor Domes™?" questions Bob.

"Isn't that ironic," agrees the consultant. "I evangelize about a structure that withstands Category 5 hurricane winds and waves, T4 tornadoes and mid-level Richter scale earthquakes developed close to hundred years ago at the Massachusetts Institute of Technology – MIT – and the building code personnel in this county only approve RECTANGULAR and SQUARE structures that have been proven to NOT withstand the forces of nature mentioned above.

"The chief inspector says I can build them as cabanas, gazebos, outdoor entertainment centers BUT NOT homes! Thousands of other counties across America have DOMES as homes and even several other counties in western North Carolina have them.

"Plus the approved square and rectangular homes costs way more than a Raptor Dome™ home of the same size. I was asked to develop a potential project in the downtown Atlanta area using these Raptor Domes™ for low income housing.

"The developer wanted to know how much a turnkey 32 foot diameter Raptor Dome™ home would cost. Of course, I explained that I had no idea what their labor cost would be; however this 800+ square foot turnkey dome with appliances and two bedrooms, a Jack and Jill bathroom and open concept kitchen/dining/living rooms would be approximately $75,000."

"Are you kidding?" questions Bob. "You can't even get a double wide for that!"

"So right, Bob," agrees the consultant. "I forgot to say that also included a 6 inch concrete foundation with radiant heated floors."

"Oh how I would love radiant heated floors!" exclaims Dorothy.

"Okay, you have sold us," states Bob. "What is the next step?"

"First, we will need to determine that the county you are moving to in south Georgia will be more open minded about the Raptor Domes Kits™ being used as homes," explains the consultant. "He or she may require you to submit architectural drawings done by a structural engineer before you can even start manufacturing.

"I realize the time is ticking towards when we feel the full force of this Grand Solar Minimum and all heck breaks loose. People will be scrambling to find a safe place to live and FLORIDA is going to one of the main areas they flee too.

"We all know it also has been experiencing more and more monster hurricanes and the Raptor Dome Kits™ will be in HUGE demand down there. You being in south Georgia will make it easy to ship the Raptor Dome Kits™ to Florida.

"As I said on Coast to Coast AM I plan to get a boat, which gives me the MOBILITY I discussed on the show. I had planned on a sailboat; however I was told by a friend that the carsickness I experience on winding mountain roads will be the same I experience on a sailboat. They had me start looking at cabin cruisers, which actually have more room especially in the galley area as opposed to a sailboat.

"First, I will travel on the 141 nautical miles of the St John's River basing out of most likely Green Cove Springs, Florida, where I will also have a 27 foot diameter Raptor Dome™ ready to flee to during a hurricane.

"If the Moon Wobble begins to flood the St John's River, which flows NORTH into the Atlantic Ocean, I will go further south on the St John's River."

"As I mentioned before, we are now heading to Jesup to scout out the area and hopefully while there we can see what their building code folks say about the domes as a potential living space," shares Bob.

"Good idea," replies the consultant. "In my books I talk about the importance of scouting out an area you are potentially looking at to live.

"You want to see how many restaurants they have, schools, hospital and physicians, shopping centers and stores like Walmart, grocery stores, hopefully a farmers market, building supply stores like Home Depot, and recreational activities.

"I would get those free real estate magazines along with a local paper. In another words – DO YOUR HOMEWORK BEFORE YOU SIGN THE DOTTED LINE."

"You are definitely on my wave length," confirms Dorothy.

"WOW, definitely a great deal to explore BEFORE the Grand Solar Minimum really kicks in," replies Bob.

"Yes, and NOT a lot of time left before that occurs," says the consultant. "If Dr. Zharkova's algorithm is functioning as it should be, then from NOW on we are going to see colder and colder temperatures even at this time of year. It is already happening here and farmers are very concerned for their crops."

"We are already seeing low 50s by 6 in the evening," shares Andy. "This new replacement meteorologist is still saying this is just a TEMPORARY situation unlike the previous two meteorologists, who shared about the Grand Solar Minimum and were replaced the next night."

"Is that right," says the consultant. "No surprise there. They sent two former NASA lapdogs to try to discredit what I was saying about the Grand Solar Minimum and telling my YouTube followers that Dr. Zharkova is basically a crackpot lowly employed professor at unknown UK university.

"Of course the Royal Society of Astronomers in the UK invited her to present a paper on the solar activities around the Grand Solar Minimum. I'm sure they invite 'crackpots' to speak!"

Everybody gives an affirmative laugh at what the consultant says and agrees to stay in touch with him.

CHAPTER 18

July 2, 2025 Franklin, NC

The family prepares for their morning ride to Jesup, Georgia for their 4th of July celebration and more importantly "scouting trip" of Jesup.

Cindy is thrilled with the results of her gem mining and has carefully placed her new little rubies in her bug-out slack pack.

"The trip to Jesup is now about 300 miles and with a lunch break about 7 hours," explains Andy as he studies his smartphone.

At around 11:45 they spot the Cracker Barrel restaurant they have read about. They see the charming old styled building with the welcoming rocking chairs spread across the front porch.

"It is nice to know there is a Christian based family style restaurant still in our country," reflects Dorothy. "The menu is very inviting too."

"I sure hope they have more of these Cracker Barrels in Jesup and the other cities," says Bob, who is still remembering his country fried steak, roasted corn on the cob and especially the apple pie alamode.

"Of course we could be lucky enough to find other privately owned family restaurants as good or better," shares Dorothy.

"I like the food but right now I am focused on that cool swimming pool," says Cindy.

As they drive up to their motel, they see a bank LED thermometer saying: **93**.

"We are definitely at an area that has the usual July temperature," states Bob.

After checking into the Days Inn, they head for air conditioned rooms to quickly change into their swimsuits.

Bob is one of the first to head for the pool and cannonball into to it fortunately avoiding sunning guests. The others clap at his accomplishment, while his embarrassed family watch as they enter the pool area.

"My husband is my THIRD child," timidly laughs Dorothy as the other guests join in on the laughter.

Soon everyone is getting acquainted with each other and plan to enjoy the upcoming Fourth July celebration together.

Upon a recommendation of the front desk associate, they find a another family styled restaurant – Jones Kitchen – offering scrumptious fried chicken, bar BBQ chicken, meatloaf, fried pork chops, etc.

"Now this is my kind of eating!" exclaims Bob.

"And very reasonable," adds Dorothy.

They decide to order one order of fried and BBQ chicken, meatloaf and fried pork chops and then share them along with numerous sides.

"Gosh I had planned to save room for a homemade desert but now I'm full," offers Andy.

"See I left room for desert," shares Cindy. "Now I just have to decide what I want. Hey, Mom let's share something."

"Great idea, Cindy," agrees Dorothy.

Before leaving Cindy and Dorothy share homemade Georgia peach cobbler with vanilla ice cream and Andy and Bob share the double fudge brownie with vanilla ice cream.

After arriving back at the Days Inn, they decide it is still warm enough to dip into the pool before retiring.

CHAPTER 19

July 4, 2025 Jesup, Georgia

They are getting themselves ready for a HOT Fourth of July by putting on red, white and blue tees proclaiming July 4, 2025 with the USA flag in the center.

All their pool friends plan to attend the festivities planned by the Days Inn before they caravan to the downtown festivities beginning with a parade, picnics from food trucks parked in the area, where bands are going to be performing and culminating in a massive fireworks display as reported on the Chambers internet site.

After enjoying a great Days Inn breakfast with each table decorated with American flags and red pancakes with blueberry syrup, they head back to their rooms.

"I'm going to call and see how things are going at the farm," says Bob.

He finds out that everything is fine except that it is even colder than when they left. He tells Ed, who is staying at the farm, about the 90 degree plus days they are enjoying.

"I sure wish I was there for the Fourth instead of here, where we will be wearing JACKETS for the celebration!" exclaims Ed. "I would never have expected this to be happening. Others are getting more and more concerned about this so called 'blip in the weather pattern' that new weather lady keeps saying. Of course good ole Frank has assured us it is just that and nothing to be worried about."

"If this keeps up and more people get concerned that this is NOT a 'blip in the weather', we may find harder to sell the farm," shares Dorothy. "I will call Jane tomorrow to see if she can get the appraisal done quickly along with photos and drone shots. In order not to panic Jane I will just say my relatives are so glad to see us that we are planning on staying longer."

Soon it is time to leave for the parade and other festivities.

CHAPTER 20

July 5, 2025 Jesup, Georgia

"That was an AWESOME COOL Fourth of July celebration!" exclaims Cindy.

"Your Mom and I have some places we need to go for several hours," explains Bob. "Andy you and Cindy can walk around the downtown area to see what Jesup has to offer."

After dropping off Andy and Cindy, Bob and Dorothy head to the county hall to find the building code department and show them the information on the Raptor Domes Kits™ the consultant has prepared.

They find Stewart, who heads up the building code department and present him the information.

"Well, this is certainly different from what I have seen before," explains Stewart. "It definitely is intriguing and the benefits are phenomenal. I am going to make copies of the material and see what my personnel think about the Raptor Dome Kits™.

"I will probably need for you to supply a strength analysis as that friend of yours that owns the business in North Carolina said.

"Each state that you do business with in the future will require it especially if it is to be a living quarters. Currently I like the building code director in Franklin have no problem with these Raptor Dome Kits™ being used for other purposes like gazebos, barns, sheds, etc. as long as they are not living quarters for humans."

"That is very encouraging news," says Bob along with nodding agreement from Dorothy. "So far we definitely like the Jesup area"

I'm sure our Chamber would be thrilled to have your manufacturing facility here and so would I," adds Stewart.

"Well that is encouraging," says Dorothy after leaving Stewart's office. "I like how open minded he was to this new form of building."

"Me too," agrees Bob. "Let's go see what Cindy and Andy have learned on their scouting of the downtown area."

"Yes, we can have lunch while you call Ed and I call Jane," continues Dorothy. "Plus we can share what we learned with the kids."

Dorothy learns that both the appraiser and the photographer with the drone can be available even before they return home.

Bob talks to Ed, who has no problem continuing to watch the farm.

At lunch it is decided to go ahead and get involved in the Raptor Dome Kits™ manufacturing as a viable new business possibly without continuing to farm as a living.

Bob calls the consultant to tell him they are ready to move ahead with the royalty agreement. The consultant agrees to fax it.

After lunch they head to the UPS Store to pick up the fax. After studying it Bob and Dorothy both sign it.

They further checkout the Jesup area picking up the suggested newspaper and free real estate magazines.

While Bob, Andy and Cindy enjoy another swim, Dorothy studies the real estate listings.

"Since we are now in the Raptor Dome Kits™ manufacturing business, I think we should just look for land to potentially put up our own Raptor Dome™ in the future," explains Dorothy to Bob as he joins her in a pool side lounger. "Land currently is quite reasonable but as the information finally comes out about the Grand Solar Minimum, prices will probably double or even go higher."

"I agree," states Bob. "I think our days of running a farm are over and actually I am thrilled to pursue this new adventure. I think up to 5 acres hopefully with a brook or stream would be ideal. We need to make sure power is currently running to the property we select though I see us eventually going off the grid as much as possible.

"I have read about Pico Power being used in Third World countries, where they take advantage of fast running rivers that can provide electricity to entire villages. Of course we have always been interested in solar."

"I agree with your plan especially when we set up our own Raptor BioDome™ using solar," states Dorothy. "I look forward to learning more about this revolutionary water-based food growing system that also secures the food from as many of these nano plastic particles that have taken over our Planet as I read in a **National Geographic** article. These particles are landing on my vegetable garden."

"That is scary," says Bob.

Soon they are joined by Andy and Cindy.

"We are going to be heading back home by way of several other Georgia cities that were mentioned by the consultant to see – Douglas, Waycross and Thomasville," explains Bob. "While we visit with their building code personnel, I want you and Cindy to scout the downtown area and get the newspaper and those free real estate magazines. Right now, let's find a place to eat!"

CHAPTER 21

July 5, 2025 Pentagon Bunker

All the members file in and are quickly seated.

The Joints Chief Admiral Hutchinson opens the meeting.

"I have spoken to the President about the current situation and briefed him on the Grand Solar Minimum we are in the midst of. He is concerned about this potential impending disaster but smartly does not want to cause chaos and panic with the 83,000,000+ Americans that are going to be greatly effected by it. So he has decided to allow US more time prepare for it. He will advise NOOA to tell all their meteorologists that there is definitely an 'anomaly' occurring but it should not be of too much concern."

"Boy, will the media have a field day with that decision," adds Marine General Davidson. "They will start their own research and most likely discover the Grand Solar Minimum and see it is NOT an 'anomaly' but a reoccurring event every 350 years give it take a few years. And if they find Dr. Zharkova's website –www.solargsm.com, and how to contact her, the President and us are going to look like buffoons!

"Yes, General, I agree with you and that is why we have to get ourselves working on VIABLE solutions like the 205,000 square miles of US caves," states Admiral Hutchinson as she turns to National Guard General Jackson and Army General Smith.

General Smith speaks up.

Yes, I have had extensive meetings with my Army Corps of Engineering staff, who agree these caves are viable options and the only problems are asking the critters to vacate.

"Excellent, General," replies the Madame Chairwoman. "Now, General Jackson, are you in the process of developing a plan that will hopefully relocate these people?"

"Yes, Admiral, and General Smith has been very helpful in regards to supplementing troops from Fort Campbell to help our National Guard like they did at the 1996 Summer Olympics," explains General Jackson. "As everyone here realizes this is going to be a monumental task moving the 83,000,000+ Americans to safety without extensive rioting."

"Absolutely, General," agrees Admiral Hutchinson. "You and General Smith are doing an excellent job as I will report to the President. Now Admiral Stetson, how is your project coming along in regards to renovating mothballed aircraft carriers into living spaces?"

"I am happy to report that the Admiral in charge of these aircraft carriers says it is also doable," explains Admiral Stetson. "These older aircraft will have to re-fitted."

"Excellent, Admiral," replies the Madame Chairwoman.

"So that leaves you General Davidson," says Admiral Hutchinson.

"I have a plan already in place to move all US embassy personnel and families out across the globe using the existing Marines at each embassy backed by my Recon units," reports General Davidson. "My other Marines can deployed and help General Jackson with his National Guard units."

"Excellent, General," replies Admiral Hutchinson. "Again I am very pleased with the progress we have made so far and I am sure the President will be also.

They all agree and adjourn.

CHAPTER 22

July 8, 2025 Topeka, Kansas

The family return home, thank Ed, and prepare their future plans at dinner.

"Well we know we are leaving here as soon as the farm sells," begins Dorothy. "Our real estate agent Jane believes it will sell fast as it is already up on the Century 21 national network. Your Dad may have a buyer interested from his morning group. Therefore we MUST keep the place as clean and neat as possible. When there is a potential buyer looking, Jane wants us ALL off the property. It will most likely be during the weekend so we will have to find something to do like a trip to a movie."

"I can't wait for the weekend to come so we can go see that new Disney film!" exclaims Cindy.

"Cindy it is now OK to tell your friends that we are moving; however, don't tell them the REAL reason we are moving," says her Mom. "You can say that we have relatives in Georgia, who want us to come live down there. Don't mention anything about the Grand Solar Minimum or our new business. Keep some mystery about the impending move."

"Sure, Mom and I like the part of being mysterious concerning the move," agrees Cindy.

Let's all retreat to our bedrooms and make sure everything is clean and neat," shares Dorothy.

CHAPTER 23

July 9, 2025 Topeka, Kansas

The next morning Bob heads to his morning meeting planning to "hook" Frank with his news.

"So Bob, tells about your trip south to Georgia," leads off Frank.

"Well it certainly was warmer there and we spent a great deal of time in the motel pool," chuckles Bob.

There are groans from everyone else including Frank.

"We have definitely decided to move to Georgia to be close to Dorothy's relatives," explains Bob. "Century 21 has already listed the farm through their national and international network. It has been appraised at $1,250,000. Jane our real estate agent believes it will sell in 30 days."

"WOW!" exclaims Ed. "May be I better have your agent look at my place though it certainly does not have the acreage yours does."

"I think that might be a little high," says Frank. "Of course as I have said before I might be interested."

(Gottcha just where I want you!" thinks Bob.)

"You should schedule appointment with Jane our agent to tour the place, Frank," shares Bob.

"Yes, I will check with Helen to see, when we can do that," says Frank.

The guys discuss the continuing cold weather and then leave Chick-fil-A.

Bob calls Dorothy on the phone to bring her up to date about the impending tour with Frank and Helen.

"I'm meeting with the sewing club, where Helen is also a member," explains Dorothy. "I will give her Jane's number. I also plan to call several real estate firms in the cities we visited and learn more about the listings we are interested in."

"Good," replies Bob. "I'm going to call down to the towns and talk to the businesses that might be interested in doing the outsourcing of the Raptor Dome Kits™ struts and hubs with their CNC routers. Andy is working with the consultant to set up our website for the business."

That evening during dinner they all discuss what has happened that day. Dorothy has spoken to the Georgia real estate agents. Bob has spoken with interested outsourcing companies. Andy has conferred with the consultant about what should be in their website. Cindy shares the reactions of her friends to the news of the impending move.

CHAPTER 24

July 10, 2025 Jacksonville, Florida

The consultant has decided to take his own road trip back to his home town. He is going to stay with his cousin at her home on Doctors Inlet right off the St John's River.

He plans to visit his old family home he grew up in the Riverside Avondale area, where his mother's father built numerous homes there. He also plans a trip to Bolles School, where he graduated from in 1964. Back then they could hardly field a football team but today they have dozens of championships and have produced several NFL players including Mack Jones formerly quarterback for Alabama and now quarterback for the New England Patriots.

Baseball is another winning sport, where retired surely to be enshrined at Cooperstown, former Atlanta Braves player Chipper Jones played his high school ball.

Academically Bolles has always been strong but now ranks in the top 10 prep schools in the USA with students from over 50 countries attending. The most exciting addition to the four campus K-12 operation is the new Center for Innovation that when finished will rival university campuses!

Finally he plans a visit to his former Post Newsweek TV Group operation – WJXT. Now a FOX station, this is where he interned and eventually became an investigative reporter covering Civil Rights, the 60s drug scene, consumer affairs and entertainment. In 1970 he left and went to Florida State University to complete his degree.

He plans to look at several used cabin cruisers in the area he has found in **Yachting World.** As mentioned previously, he had originally looked at a sailboat and he knows quite a bit about buying a used sailboat from doing his research for authoring **Sea Gypsy Live-Aboards.**

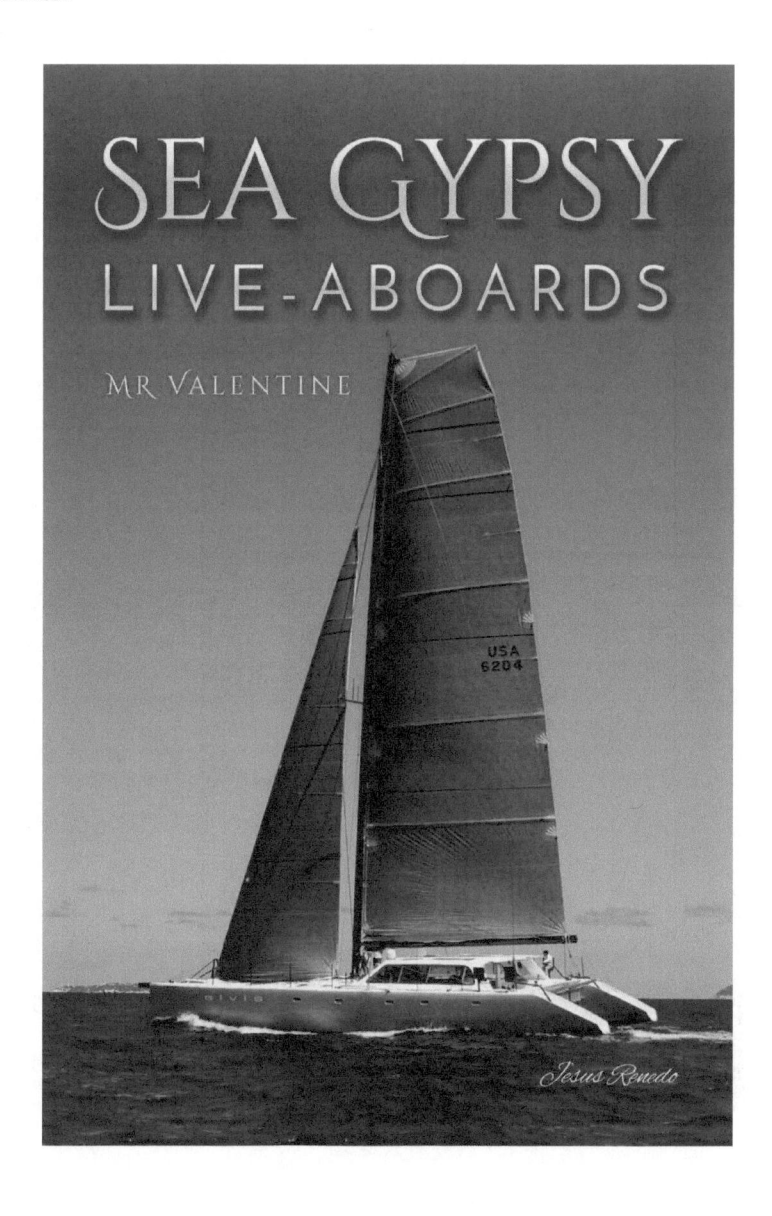

Marinas can be excellent places to look for "don't wants" as the old saying is: "The BEST day was when I bought the boat; however, the BETTER day is when I SOLD it!"

The consultant knows this going into this adventure; however he plans to live on it for the rest of his life and run his numerous businesses from it. One business he sees growing and growing will be consulting people, who finally hear of the Grand Solar Minimum from his appearances on radio and TV and his books.

They will be desperate to relocate paying PREMIUM fees! At least his Raptor Domes Kits™ USA business will be in capable hands; however he sees other interested parties all over the world wanting to do the same royalties deal.

(As of the publication of this book the USA business along with other countries is still available and you can learn more at http://www.raptordomes.com.)

CHAPTER 25

July 11, 2025 Topeka, Kansas

"Good news, Dorothy!" relays an excited Jane. "We have already had an offer for your farm from several parties including a person from Dubai. He is has offered the closest to the asking price and your friends Frank and Helen have put in a bid of only $850,000.

"This is INCREDIBLE!" exclaims a stunned Dorothy. "What is the Dubai offer?"

"One million dollars," replies Jane. "He will be sending his representative to see the property this week. I will meet his private jet and hire a limo to take us to the farm. Don't worry we gladly will pick up the tab for the limo."

"I will start cleaning like a mad woman and inspect every inch of the place and have Bob and Andy do the same with the barn," says Dorothy.

"Excellent," replies Jane. "I will keep you updated. By the way the potential owner is not from Dubai but has chosen to live there for business purposes."

Dorothy does not know how she is going to keep from telling this exciting news until dinner but goes immediately to cleaning.

Finally dinner time arrives. Cindy shares about her day in summer school and how her friends keep asking about her impending move. Andy discusses his progress on their new website.

Dorothy realizes this is the PERFECT time to share her news.

"Well I don't know if I can top Cindy and Andy's recap but I will try," replies Dorothy. "I got a call from Jane and as you predicted, Helen and Ed have submitted an offer."

Bob interrupts: "I knew he would! He has wanted our place for years!"

"Unfortunately his offer is only $850,000," continues Dorothy.

"I should have figured he would do that," replies a dejected Bob.

"However I did not mention there is another offer," continues Dorothy.

"I sure hope it is above Frank's or we might have to set our price lower," shares Bob.

"I think it is somewhat higher – like ONE MILLION DOLLARS!" exclaims Dorothy.

"What! ONE MILLION DOLLARS!" exclaims Bob and Andy.

"Yes, and the REPRESENTATIVE of the potential buyer is privately flying in this week and Jane and her company are picking up the tab for a limo!" exclaims Dorothy. "The potential buyer lives in Dubai."

"Is this a dream I am experiencing?" questions a stunned Bob.

"No, and after dinner I want you and Andy to start cleaning up the barn," says Dorothy. "Cindy, you and I are going to work on your room and the laundry room. Andy you will be also cleaning up your room."

"Come on Andy, we have our work cut out for us," replies Bob as they both head for the barn.

"Mom, where is 'Dubai'?" questions Cindy. "I have never heard of it."

"Good question, Cindy," says Dorothy. "And one you and I can look up on the NET after we finish our chores."

Before retiring everyone meets to pray over the upcoming visit.

CHAPTER 26

July 16, 2025 Topeka, Kansas

The visit is over and the family await a report from Jane on the outcome. Finally it comes as everyone gathers and listens over the speakerphone of Dorothy's smartphone.

"Well the visit took place as I shared," begins Jane. "He flew in on a Gulfstream and I met him along with owner of the agency.

"He was not Arabic as I had expected but American and a former US Navy SEAL. He explained that the potential buyer and he had been childhood buddies in Texas. He chose the Navy and the other buddy chose entrepreneurship. Both had their adventures; however his buddy had become a multi-millionaire.

"They continued to stay in touch throughout the years. The former SEAL wound up in the security business in Dubai and told his buddy he should consider moving to Dubai as he now knew a number of wealthy Arabs living there.

"The move was made and both have prospered; however the potential buyer wants to have a place like your farm that he saw on the NET. So he sent his best bud to represent him."

"WOW!" exclaims Dorothy. "This is like a fairy book story. Two best friends become successful and we are involved in the story."

"You are especially going to like that the SEAL saw the cross in the living room and said both he and his family and the potential buyer, who is a long time bachelor, are devout Christians," states Jane. "They both have been since high school and currently lead a Bible study in Dubai."

"We prayed before he visited that potentially the right person or family would eventually have our farm," replies Dorothy.

"Well, that is definitely in the cards if you decide to except an offer of $1,250,000!" exclaims Jane.

"DONE!" Bob and Dorothy say in unison with Andy and Cindy joining in.

"I sorter thought you would accept this offer, chuckles Jane. "I will immediately draw up the papers for the sale. I have confirmed that the potential buyer has the available funds. You will sign and then I will fax over the contract for his signature."

After the call there is high fiving and most importantly prayer.

Jane has said they have plenty of time to move out allowing for Andy and Cindy to finish summer school. Bob also is relieved to find the buyer wants the livestock as agreed in the contract.

"I can't wait to share the good news tomorrow morning with especially Frank," says Bob.

"Please don't gloat and rub it in, Bob," warns a stern Dorothy. "That is NOT what Jesus would do."

"I won't…I promise," replies Bob.

CHAPTER 27

July 17, 2025 Topeka, Kansas

Dorothy reminds her "third child" Bob not to gloat over Frank.

Bob arrives with deadpan look at the local Chick-fil-A and greets everyone.

"Well, I guess by now your agent Jane has relayed our generous offer for your farm," immediately starts Frank.

"Yes, she has, Frank, and Dorothy and I appreciate it," replies Bob. "Unfortunately another party outbid you."

Everyone else in the group are continuously turning their heads between Frank and Bob.

"Well, I know if it is a few thousand over our $850,000 you will choose us," says a smug Frank as he emphasizes the amount.

"Yes, Frank, we would have chosen you and Helen if that was the case," replies Bob.

The rest of the heads turn towards Frank.

"You are saying it was more than a little over $850,000?" questions Frank.

"Yes it was **$400,000** over yours," explains a calm Bob.

"WOW!" exclaims the others except a stunned Frank.

"You are kidding, right?" questions Frank.

"Nope and Dorothy and I signed the contract," states Bob firmly. "We most likely will never meet the buyer as he sent a representative on a Gulfstream to see and buy for **$1,250,000**."

"Oh, God, I hope it isn't some foreigner especially Chinese or from the Middle East," replies Frank with the others bobbing their heads in agreement.

"No he is a Texan, who happens to live abroad," assures Bob to Frank and the others.

"I guess we can deal with a Texan," replies Frank as he looks for approval from the others. Well I just wanted to hear what happened but I have to leave now."

After Frank leaves, the rest of the group congratulate Bob and buy him breakfast.

After they all leave, Bob immediately dials up Dorothy to share the blow by blow event.

"I am proud of the way you handled it with that blowhard!" assures Dorothy. "We have lots to do in a short time including deciding where to live – Jesup, Waycross, Douglas or Thomasville."

CHAPTER 28

July 18, 2025 Pentagon Bunker

They members are all seated awaiting the arrival of the President himself for the briefing. The Admiral Chairwoman has even relinquished her seat to the President, who is arriving less his Secret Service escorts.

All rise to face the President, who salutes and motions for them to be seated.

"Thank you for inviting me to this briefing, Madame Chairwoman," says the President.

"Sir, it is an honor to have you here," replies the Chairwoman. General Jackson, I am going to have you ramrod this briefing."

"Thank you Admiral and welcome Mister President," states the head of the National Guard. "I have been working very closely with all the members and especially General Smith of the Army. As you know Mister President it is my duty to protect our USA based citizens from all catastrophic events like this Grand Solar Minimum.

"We estimate that over **83,000,000** Americans are going to effected by this potential 35 year event. We have been studying every possible way to safely place as many of these Americans as possible in over 205,000 square miles of caves, in abandoned mines, in the bottom of subway stations in cities like New York and Atlanta, in huge underground basements of buildings, and even on mothballed aircraft carriers.

"The Army Corps of Engineers has been extremely helpful with studying, which cave systems are feasible for this long term event and I am happy to report well over 75% can be used. As for the mines, we know that a number of them have been readied for you and your family, the Vice President, members of Congress and the Senate, Secretaries, Pentagon officials, etc.

"Hopefully we can locate other safe mines that companies still own and operate and re-outfit them. We have talked to these mega building owners about how we want to have their basements ready for their families and the families of the occupants. We have not mentioned why but only told them this is a part of a new disaster plan. Ditto for subway operations. As for aircraft carriers, I will let Admiral Stetson speak to that later."

"As for the handling of potential chaos, we are working very closely with General Smith and General Davidson. It is possible that local police will be not available as we have seen before. As you are aware there has been great friction between citizens and police over many issues. They are reporting through their representatives as not feeling respected for constantly putting their lives on the line. So unfortunately we can't count on them."

"Yes, this greatly concerns me too, General," states the President. "This is not the time for them to abandon their jobs and we must assure them they are needed. I plan to stress this in my speech to the nation."

"Thank you, Mister President," replies a relieved General Jackson. "As I mentioned, Generals Smith and Davidson are going to use their personnel to help with the problems expected to be encountered.

"General Smith has assured me that all of his crowd control trained personnel from Fort Campbell will be available as they were during the 1996 Summer Olympics. General Davidson's personnel will principally be protecting our embassies around the world; however the

Marine Recon can help with the personnel at Fort Campbell. So can Army Rangers, Navy SEALs, and Air Force Para-Rescue.

"Of course you as President will have to declare martial law allowing for US military personnel to operate in these capacities."

"So true, General and I will be announcing this during my speech to the nation," states the President.

"At this time, I would like to have Admiral Stetson address the usage of mothballed aircraft carriers," concludes General Jackson.

"Thank you, General and Mister President," replies Admiral Stetson. "We actually were made aware of this possibility by a long time Preparedness and Survival Skills consultant having written eight books on the niche.

"I went to the Admiral in charge of the US Navy mothballed fleet and asked him the viability of this. He agreed it was doable. I asked how many of these aircraft carriers were available that could be re-configured. He said of the twenty available most likely fifteen were re-configurable.

"The larger ones use to carry one thousand personnel. These would have to be re-configured to handle families. The mess hall should easily handle the dining requirements. The hospitals can be restocked and updated. The upper deck can be used to grow food with a revolutionary water-based system that is easily constructed and maintained.

"All this being said each carrier will be able to accommodate probably 900 citizens and crew. I know this is a drop in the massive bucket but it is doable."

"Interesting solution presented by this expert," says the President.

"In addition, we have a number of bases on the Atlantic, Pacific and Gulf Coast that can right now accommodate possibly more citizens along with the regular personnel and their families," adds Admiral Stetson.

"Thank you, Admiral," says Admiral Hutchinson. "I would like to have General Alexander, US Air Force, brief us on what he and his personnel can do."

"Thank you, Admiral and welcome Mister President," replies the General. "As you all are aware we also have numerous bases that can be used. Also we have mothballed aircraft that could actually be converted to living space for at least one family.

"I'm sure all the aforementioned services have the same situation. I know that one hedge funder took one of our abandoned missile silos and re-configured it into several luxury condos that were sold almost immediately. I would have to check to see if more available."

"Thank you, General Alexander," states Admiral Hutchinson. "Finally there is Admiral Jones of the Coast Guard.

Thank you and it is an honor to have you here Mister President," replies the Admiral. "Of course we are responsible for protecting our American coasts and waterways. I believe that you are going to see people taking their boats and heading to particularly South Georgia and Florida.

"This could create the equivalency of a rush hour on a LA expressway. Plus boaters from foreign countries including undesirables could flood in.

"Fortunately we are going to have enough vessels and personnel to help the governors of Georgia, Florida, Alabama, Mississippi, Louisiana and Texas with this potential chaos. Yes, not every vessel will be detected but we certainly will get most of them. Of course it will be nice to have the Navy and Air Force helping us if they are available."

Both General Davidson and Admiral Stetson acknowledge this with a thumbs up.

"So Mister President you can see these men and women have been busy around the clock preparing plans and putting them into action since we learned of this Grand Solar Minimum," continues Admiral Hutchinson. "I am sure you have questions that we can hopefully answer."

"Thank you Madame Chairwoman and all of you gathered here today," begins the President. "Yes, it is possibly the most monumental task the United States has seen since World War Two. This time our enemy is coming ashore on American soil.

"This time it has over **83,000,000** Americans as its target. However like before we will meet this enemy with every available person in our Armed Forces. I plan to go before the American people like Roosevelt did after Pearl Harbor and tell them the FACTS about the Grand Solar Minimum.

"I will have to declare martial law, as much as I hate to do so. I have decided to do this on Labor Day weekend in order to give you all more time to have made more progress. I plan to continue to use the folks at NOAA to use the same story of a "freak" weather event, which the Grand Solar Minimum actually could be considered. I will allow them to tie into global warming."

The President immediately departs and the others follow him out.

CHAPTER 29

August 1, 2025 Topeka, Kansas

It is getting closer to moving day scheduled for August 14th. Andy and Cindy are out of summer school. Dorothy is getting both of them involved in packing their own things they want to take and setting aside things that can be given away. Bob is making sure all the livestock are caught up on their vaccinations along with their two cats and dog. He and Andy have been repairing things that have been put off.

The weather continues to get colder but Frank and the others still believe Janice and her story of how global warming is causing this freak event. Of course before this, Frank had thought global warming was completely nuts.

Also Andy has continued to work with the consultant on preparing their new website. The consultant has agreed to help them with marketing and developing a press release.

Bob has located a business that wants to get their CNC router involved in manufacturing the struts and hubs.

They have made the decision to move to Jesup. They like the town and its schools, hospital and shopping. The Chamber Executive Director Mary is thrilled with having the Raptor Domes Kits™ business and that Bob and Dorothy have already joined the Chamber.

Cindy is thrilled that she can continue her gymnastics training and Andy is looking forward to his senior year there.

Mary, who also sews, plans to introduce Dorothy to her sewing club.

They are all excited about the new land they will be moving too. The seven acres has a fast flowing stream and plenty of sunlight. They have decided to rent a RV instead of build, which is thrilling to Cindy. They are planning to use the first manufactured Raptor Dome Kit™ to build a Raptor BioDome™ powered by solar panels. The second one will be a cabana for the swimming pool like the one seen here with a Raptor BioDome™.

The third will be as a place for Andy and Bob to store their tools they will be bringing along with items they won't be able to get into the RV.

The fourth one will be a gymnastics area for Cindy and her friends along with a family workout area complete with sauna.

The fifth will be a she dome for Dorothy with her sewing machine along with a laundry area so she can hear the washing machine and dryer.

The best use of the Raptor Domes™ will be as models for people to see on their website and in person.

Stewart, the head of building code department has approved the Raptor Domes™ for these purposes and he is further studying about the dodecahedron shape and how domes in general have been approved as living spaces in hundreds of other counties across the USA.

CHAPTER 30

August 14, 2025 Topeka, Kansas

"That was a wonderful "Going Away" party, Dorothy," says Bob as they continue to prepare for their move tomorrow.

"Yes, it was and Jane sure knows how to organize one," replies Dorothy. "She invited my sewing club members, your farm buddies, Andy and Cindy's friends and even the entire staff at Century 21."

"Yeah, even Frank attended with Helen and remained civil the entire time," remembers Bob. "I do still think he was hoping our deal with the Texan would fall through."

"Speaking of the Texan, his former SEAL buddy is flying in tomorrow to get everything ready for his friend and should be at the farm tomorrow according to Jane," shares Dorothy. "We may be here just getting everyone in the RV when he comes to the farm. He will oversee the installment of the solar panels like we had planned to use."

"If he does come by, maybe he can give me a contact company for our installation in Jesup," replies Bob.

"Good idea and in the meantime I have to make sure everything is ready for the movers," answers Dorothy.

"I'm going to review the manual and DVD that came with the RV," says Bob. "I think I will take Andy and Cindy on another test drive too and stop off at DQ as going into that parking lot should be a REAL test of my skills. I will bring you your Lime Freeze."

The test drive is a success and back on the farm they all enjoy their Dairy Queen treats.

CHAPTER 31

August 15, 2025 Topeka, Kansas

It is moving day and the movers arrive right on time to begin the process.

Dorothy is in charge of the move, while Cindy, Andy and Bob load up the RV.

The night before Bob gassed up the RV so they could proceed immediately on their way.

Dorothy and Bob have decided to stay at the Day's Inn upon arriving in Jesup. The moving company told them that it would be impossible to get everything into their storage unit upon arrival.

About noon Jane appears at the farm with the former US Navy SEAL. The last items are being placed in the moving van and the RV is ready to go.

"Glad I got a chance to see you before you left, says the representative. "My friend and I wish you the best in your adventure especially your new Raptor Domes Kits™ business that Jane shared with me. I have been a fan of domes for years after reading about Bucky Fuller. I think you are on to something that will only get more and more attention especially, where there are hurricanes like Florida."

"Jane told us you are here to oversee the installation of the solar panels," says Bob. "Got any tips as we also plan to install them on our new property?"

"One thing I have learned is if you do put them on your roof, you had better make sure they are tightly secured. Here with all the tornadoes I plan to place them on the ground. Yes, they could be wiped out with a direct hit; however the slightest winds from even a thunderstorm could lift them off the roof. Another advantage of having them on the ground is the ability to have them automatically adjust with the sun."

"Yes, I agree and have you got a contact for company that you are using?" questions Bob.

"Sure do and they handle installations in the USA and Dubai as that is where I learned about them for my home and then my buddy's home," says the former SEAL.

Soon the last item is on the van and the family follow it out their driveway for the last time.

It is late when they arrive at the motel but still warm enough to get into an inviting pool before retiring.

CHAPTER 32

August 16, 2025 Jesup, Georgia

After a hearty breakfast it is time to meet the movers at the storage unit.

By noon everything has been stored away and the family head to another private restaurant recommended by the front desk – Captain Joe's Seafood – just on the outskirts of Jesup.

They have decided to stay one more night at the motel and enjoy a leisurely time by the pool.

Later they head for the local Walmart to stock up on supplies and then it is off to another restaurant – One Love Island & Soul Food Restaurant, which specializes in Caribbean with jerk chicken, Caribbean meatballs, coconut curry shrimp, conch in butter sauce, etc.

A late swim tops off the evening before retiring.

CHAPTER 33

August 17, 2025 Pentagon Bunker

Everything has come unglued from the planned announcement on Labor Day weekend by the President after a well known conservative talk show host with a huge following has delved into what he is calling "The Grand Solar Minimum Cover-up" causing the phones of Representatives and Senators and even governors to go into overdrive.

The aides are at a loss for answers and start to replay the show featuring Dr. Zharkova and the Preparedness and Survival Skills consultant. It appears the talk show host learned of the Coast to Coast AM appearance and decided to do more research. He and his staff found YouTube video after video featuring Dr. Zharkova.

All the members are in a somber mood awaiting the arrival of Admiral Hutchinson, who is coming directly from the President's office.

The Admiral arrives also in a somber mood and takes her seat.

"Well you would have to be living under a rock not to have heard the firestorm generated on that talk show with Dr. Zharkova as a guest along with that consultant," starts off the Madame Chairwoman.

"Needless to say governors, Representatives, and Senators have lit up the White House phones along with media representatives. Now the President plans to hold a televised press conference tonight."

"I knew this was going to happen and mentioned it in a previous meeting," states General Davidson of the Marines. "I am sure it is going to have repercussions at our embassies so I had better alert our personnel."

"Yes, General; however alert them AFTER the President's first mention of the effects of the Grand Solar Minimum," replies the Admiral. "Ditto for your Marine Recon staff. You are excused to begin all these preparations."

General Davidson rises salutes and exits.

One by one the other members of the Joints Chiefs makes their speech to the Chairwoman, gets her reply, salutes and leaves.

All have massive preparations ahead of the President's televised press conference.

CHAPTER 34

August 17, 2025 Jesup, Georgia

When the family enter the Day's Inn restaurant everyone is talking about the Grand Solar Minimum!

They soon learn from a couple seated next to them about how the consultant and Dr. Zharkova have been featured on a conservative talk show that has resulted in the President calling a nationally televised press conference tonight at 8 PM EDT.

They listen but do NOT share that they have known about it for quite some time.

After breakfast, they immediately head to their rooms to get everything back in the RV.

They decide to head back to Walmart, where they find a crowd like during COVID buying loads of food items, paper goods, canned goods and toilet paper. They follow suit being glad they have an air conditioned storage room. They plan to see if they can get another unit.

Also because the funds from their closing have cleared the bank they decide to also rent a two bedroom apartment if they are available.

Dorothy calls Mary at the Chamber (which they are so glad they joined!) to see if she knows of place to rent. Fortunately she does and calls the manager of the apartments, who is also a Chamber member and a sewing club member. It is arranged for the family to come over immediately as they have a vacancy.

They arrive to find a nicely kept small new complex and a very friendly Sue that shows them the three bedroom unit. It is very spacious and features the open floor concept with modern kitchen appliances and plenty of storage space. The master suite has its own bathroom and the other two bedrooms share a Jack and Jill bathroom.

Cindy is thrilled to see a pool!

The family decide to take the unit with a 6 month lease with option to renew. The papers are signed and a set of keys is presented.

Dorothy starts to immediately unload the new supplies from Walmart and everything on the RV is unloaded by Bob, Andy and Cindy. Then they head to the storage unit to get everything they can get into the RV. Bob calls to get a U-Haul so that he and Andy can move the stored furniture into the apartment.

By five o'clock everything is in place and they take a break to use the new pool and meet neighbors. Everyone is talking about the Grand Solar Minimum and wondering if it is going to effect them.

Again Dorothy and Bob stay quiet about their knowledge of the Grand Solar Minimum and use the story about relatives in the area, who wanted them to move here.

While getting ready for their first new home dinner it's emphasized the importance of staying quiet about the REAL reason they have moved to Jesup.

Thank goodness their big screen TV immediately works and they await the President's press conference.

CHAPTER 35

August 17, 2025 The White House

Members of the electronic, print and internet press representing not only the USA but countries around the globe are seated waiting for the President to make his appearance.

Soon he arrives as the members stand and are seated. They have been told in advance how the presentation will take place with numerous slides, videos and most importantly others involved in preparation for the Grand Solar Minimum.

The President does not waste any time immediately introducing an expert from NOAA who explains what a Grand Solar Minimum is with slides. Next the screen goes live with Dr. Zharkova, who is introduced as a renowned solar physicist and expert on the Grand Solar Minimum.

Next is Admiral Hutchinson, Chairwoman of the Joints Chief, who turns it over to National Guard General Jackson. He discusses what is currently being done in regards to the caves, underground subway areas, large building basements, and abandoned mines. He then introduces US Army General Jackson, who discusses how the Corps of Engineering is involved.

Then comes Marine General Davidson, who discusses security precautions that as he speaks are being undertaken at all US embassies to protect and if necessary evacuate American personnel and their families.

The President makes a short speech thanking those present for their efforts and giving them continuing support as he then says how he has decided to immediately instigate martial law in the USA. The press corps all gasp at the announcement.

Now the President opens it up for questions that come pouring in like an avalanche!

Two hours later a noticeably exhausted President retreats from the press conference as media reps start talking on their cell phones and rushing to do the traditional stand-ups outside the White House, where numerous ones first express how cold it is for an August night.

CHAPTER 36

August 17, 2025 Jesup, Georgia

Bob, Dorothy and Andy are stunned by the President's announcement of martial law; however so glad and blessed by GOD to have escaped from Topeka, Kansas!

"I feel somewhat guilty knowing what we thought we knew from listening to Dr. Zharkova's video and what the consultant has said selling our property to the Texan," says Dorothy.

"I too feel that but we must realize that we did NOT DEFINITELY know it was the Grand Solar Minimum; therefore we are NOT guilty of misleading him," emphasizes Bob. "I think now the MOST important thing on our plate are the steps we need to take immediately."

"First, we must get Andy and Cindy into the Jesup school system tomorrow if we are allowed to move around," replies Dorothy.

"Totally agree and I plan to contact Stewart to see what impact the declaration of martial law will have on our new Raptor Domes Kits™ business," adds Bob. "It might speed up approval for housing as there is going to be a mass exodus from those states affected by the Arctic temperatures of the Grand Solar Minimum."

"Yes, it could speed up our business if we are allowed to move around and have the struts and hubs manufactured on the CNC router," adds Andy.

"Andy see if you can reach the consultant and hopefully he will have some suggestions," says Bob.

"Hey while you all are doing this can I spend time in the pool?" asks Cindy.

"Sure, after you help me with things we need to do here," assures her Mom. "We have a busy day ahead; however let's pray together first thanking GOD for protecting us and getting us into our new apartment and SECOND and most importantly FUTURE GUIDANCE!"

They hold hands together in a circle with Bob leading the prayers.

CHAPTER 37

August 31, 2025 Jesup, Georgia

The declaration of martial law has had way less effect in Jesup than more crowded areas like Atlanta. The governor of Georgia has decided to set up extensive roadblocks with every road or expressway coming into the state. These areas are manned by Georgia State Patrol, sheriff departments and local authorities, who are turning away people fleeing south.

It is the same in Florida, Alabama, Mississippi, Louisiana and southern Texas. California has closed all ways into their state with the governor using his National Guard.

Wealthy people are booking private jets to take them to warmer places they have either purchased before or now. Mega yachts are sailing to warmer places too.

Jane, their real estate agent, calls Dorothy to see if she can find a place for her, husband, mother and the husband's family IF they can even get out of Topeka to find the number has been changed to an unpublished number.

Bob has had his phone number changed as have Andy and Cindy. This was a painful decision at the suggestion of the consultant. He has arranged for all of them to have new secure encrypted emails and his friend's encrypted secure Kryptall smartphones.

The good news is that Andy and Cindy are back in school and enjoying classes and new friends.

The other good news is that the new company website is finished. The consultant has done a press release as his former background of owning a public relations consulting business in Atlanta makes it easy to do for his new friends.

It is already getting results especially from warm area media in Florida, who know that the brunt of the hurricane season is in full swing with potentially several monster hurricanes to hit Florida like Hurricane Ian.

Stewart has decided to approve the Raptor Domes™ as living quarters. Now Bob and Andy can start building several for their own living quarters.

Yes, everything has worked out for Bob, Dorothy, Andy and Cindy with **God's Reset Phase 1**.

Epilogue

As I mentioned in the Introduction there is at least one more phase of God's Reset - Moon Wobble that will take place all over the world starting in 2032.

This event will cause coastal flooding all around the world! Ocean and Gulf waters should come into the shore areas at least another **150 FEET**, which will be devastating to areas of New York City! There is a good chance that certain subway stations will be effected. Here courtesy of **The Sun** is a map of the world and the cities that will be most effected by it.

FLOODY MOON

[Image showing a world map with numbered flood-risk locations]

USA:
1. New Orleans
2. New York City
3. Miami

AROUND THE WORLD:
4. Osaka, Japan
5. Alexandria, Egypt
6. Rio de Janeiro
7. Shanghai

As you see New Orleans is another city in the USA that will be greatly effective by it along with Miami.

The **BIGGEST** God's Reset is possible during the next few years that happens every 9,000 plus year and in the Old Testament is known as The Destroyer.

The New Agers call it the 11th Planet or Nibiru. However HERE (https://sservi.nasa.gov/articles/scientists-reject-impending-nibiru-earth-collision/) is what NASA says about it.

Research shows that this so called NON-EXISTENT object that looks like a comet with a CROSS for a tail, was the cause of Noah's flood, the parting of the waters for Moses, and other events recorded in histories of numerous countries!

By the way NASA itself showed this object between Planet Earth and the sun but pulled it down! I saw the photo from NASA myself!

So it seems GOD has definitely **TWO** Resets scheduled and potentially another one too!

If you want to be one of the LIMITED ones to work directly with me, then email me at hughsimpson@gmx.com Subject: I Want to Work With You.

You as a purchaser of this book will have the following opportunity:

1. I have created two videos on exactly what I have in both my slack pack and backpack. This is the first things you need to have.

2. If you are interested in watching these videos and then asking me questions as related to the videos, then you will pay **$100** to my PayPal account in order to be in the first group of a MAXIMUM of **10** participants on a Zoom call.

3. I will take only **10** calls at $100! So if you are number **11** you will then pay **$125** along with the other people on the second round of 10 calls of 10 participants.

4. Each successive round of 10 calls of 10 participants will go up **$25** until we reach **1000** participants! And that will be the MAXIMUM number of participants allowed!

I am only offering this to the first **1000** purchases of the book!

There will be another video where each of the 10 participants at $100 will get a chance to have a Zoom call with my very knowledgeable land real estate broker that has decades of experience selling land in the south Georgia areas mentioned in the book.

As you can imagine the price of this land will keep going up the more people read this book and decide to take ACTION!

If you opt not to watch the second video, then the first person in the second group will take your place at $100. You will not have the option to see the second video and ask questions!

As before the second group of 10 wanting to be on the call will pay $125. Third group $150. This will increase by $25 just as on the first call.

So all TOTAL you will be paying $200 to be in the first group. $250 to be in the second group. $300 to be in the third group. $350 to be in the fourth group. $400 to be in the fifth group. $450 to be in the sixth group, etc.

Of course with people dropping out of their group, others will have the opportunity to take their place at the lower price bracket.

God Bless You & Prepare YOU!

Hugh Simpson

(http://www.ihughdesign.com)

hughsimpson@gmx.com

Valentine International LLC

(http://www.valentine.international)

Raptor Domes Kits™

(http://www.raptordomes.com)

Made in United States
Troutdale, OR
02/17/2025